Advancing Physics AS

Edited by
Jon Ogborn and Mary Whitehouse

CD-ROM edited by Ian Lawrence and Mary Whitehouse

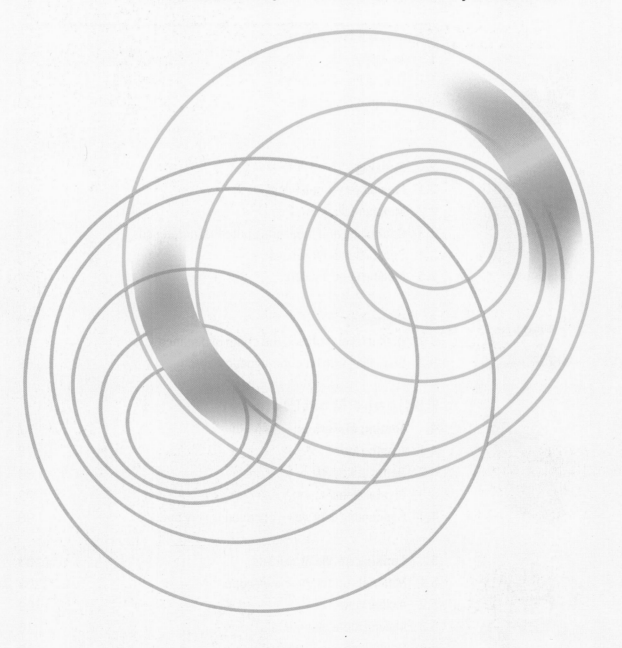

Institute of Physics Publishing
Bristol and Philadelphia

Contents

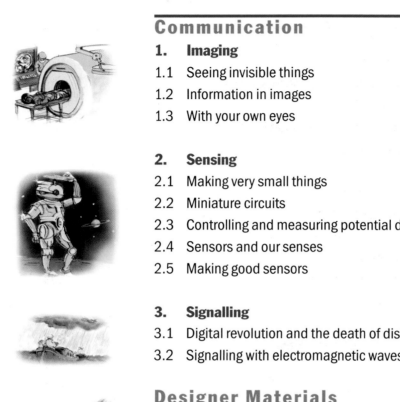

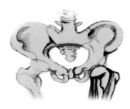

UNDERSTANDING PROCESSES

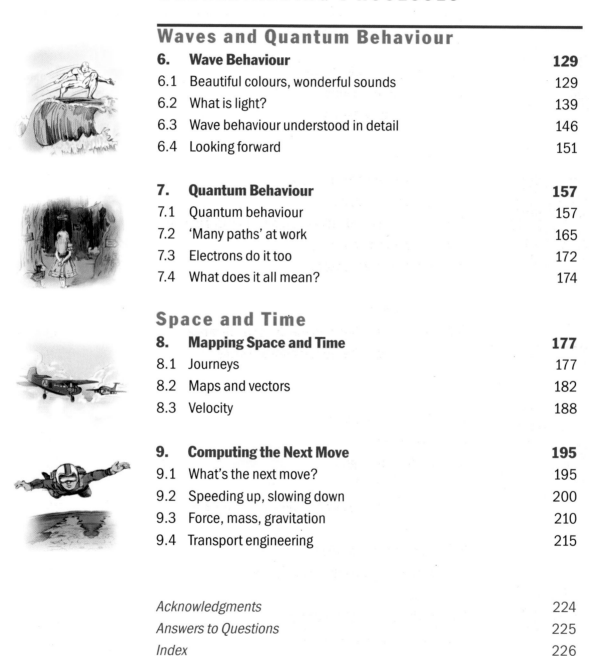

Waves and Quantum Behaviour

Space and Time

Introduction

Advancing Physics shows how physics is exciting, always developing, of great practical use in many ways and in many careers, as well as getting at some of the deepest truths about the physical world. AS level *Advancing Physics*:

- is thoroughly up-to-date
- uses Information and Communications Technology as a vital part of physics
- offers something for everyone in its variety of ways of looking at doing physics
- provides a broad picture of physics as a subject, valuable whether or not you continue further with the subject, with the accent on what you might want to do in the future
- positively encourages your own individual initiative and interest.

Advancing Physics describes essential core ideas in physics in a variety of up-to-date and interesting settings. For example, work on images recognises the crucial importance of visualisation in physics today, and shows you how electronic images are acquired and processed in for instance medical scanning and astronomy. Work on electric circuits is done as part of a study of modern instrumentation using microsensors, teaching you also how careful experiments are designed and done. Work on materials recognises that today the world is moving more and more towards 'designer materials', materials built – atom by atom if necessary – to do a particular kind of job. The study of waves looks at how the nature of light has puzzled people for centuries, and continues to do so. This leads to a very novel but very simple introduction to the strangeness of quantum ideas in physics. The study of vectors, motion, forces and gravity recognises the fundamental importance of computers in predicting, controlling and understanding motion, from air traffic control to guiding space craft. In these and other ways, *Advancing Physics* shows how ideas in physics have broad human importance, relate to a wide range of technologies, and are intellectually stimulating and satisfying.

This Student's book is written as a series of narratives, each telling a story about the physics involved from a particular point of view. Sometimes the point of view is physics in use, for example in medicine, in telecommunications, in testing and using materials, and in games and in aircraft guidance systems.

Sometimes the point of view is human and historical: how people had ideas, and what impact the ideas have had on human culture. Sometimes the point of view is new ideas themselves: how they fit together to make a satisfying structure which nevertheless challenges everyday common sense thinking. We recognise that peoples' interests vary widely; this is why the book tells its stories from different points of view.

Much of the thinking is presented visually, both in the book and on the CD-ROM. In this way, abstract ideas are made more accessible. In particular, we have made a special effort to make mathematical thinking in physics both clear and attractive.

Throughout, the aim is involve you, allowing you to develop your own ideas and interests. The book and CD-ROM contain a wealth of material from which such interests can take off, and the examination is planned to reward your initiative and independence in learning.

Specification (syllabus) details for *Advancing Physics* are included on the CD-ROM.

Links between this Student's book and the CD-ROM

MORE TO SEE ICON more to see on the CD-ROM: images, readings which develop ideas further or give more background, other things to look at

MORE TO READ ICON more to read on the CD-ROM: help with difficult points; extensions of ideas; readings giving further information or background

COMPUTER FILES ICON computer activities and files on the CD-ROM: data to work on; computer models and simulations; computer tools to analyse images and signals

A-Z ICON references to the A-Z of physics on the CD-ROM: use to look up terms; to revise ideas; to get a different point of view

How to use this book and CD-ROM

The book and the CD-ROM go together, and both are essential to the *Advancing Physics* course. The book provides, in each numbered section of each chapter:

- Narrative text telling the story of a part of physics, putting it in context, and showing you why it's worth understanding
- Essential ideas, often set out graphically, showing their logic in visual form
- Many further illustrations
- Short 'Try these' questions at the end of each section, to start you off on being able to make calculations and arguments for yourself
- 'You have learned' lists of the main points in that section
- References to useful resources on the CD-ROM and, at the end of a chapter, a summary of all the essential points and some further more extended questions to try.

The CD-ROM provides, for each chapter:

- Activities, including some using software on the CD-ROM, and many experiments to do
- Questions at different levels, from 'warm-up' and practice exercises to questions at AS examination level
- Readings which extend the material in the textbook, or provide interesting alternative views of the physics in the course, and its background, with emphasis on the people involved
- Images to look at which add to those in the textbook, and diagrams to study
- Computer files for use with a variety of computer software, notably spreadsheets and modelling packages
- Checklists of things you should have learned to be able to do and, for the whole course, a 'How to...' section helping you to acquire skills useful in all areas of physics, and an 'A-Z of physics' covering all the ideas in the course in a form useful for revision.

Studying a section of a chapter

Each chapter is divided into numbered sections. When studying a section, we suggest that you:

- First read the section all through to get a sense of what it is about and of the story it has to tell
- Go back and study the diagrams which display ideas pictorially. You will have to work at these to get their full message. Look carefully at how the logic of arguments and relations between ideas are laid out graphically. Try reconstructing the diagrams for yourself: this can teach you a lot
- Go to the CD-ROM to find further things to do, as sign-posted in the book
- Try a range of activities and questions on the CD-ROM, guided by your teacher
- Re-read the text, aiming to see how all the ideas fit together and make sense
- Work through all the 'Try these' questions at the end of the section. Use the answers to work out how to do them, if necessary!
- Look up ideas you don't understand well enough in the A-Z on the CD-ROM (or look them up anyway to get a view different from the one in the book)
- See how you are doing on the CD-ROM Student's Checklist, and the book Summary Checkup for the chapter
- Personalise your CD-ROM by creating your own 'shadow file'.

Revising for tests or examinations

- Review your 'shadow file' notes
- Browse the A-Z terms listed for each section of a chapter
- Try more questions on the CD-ROM
- Read through activities on the CD-ROM to be sure you remember what they teach you
- Stretch yourself with some further readings
- Try further activities including ones using software
- Have another read of the chapters you will be tested on.

The World Wide Web

The *Advancing Physics* Web site (http://post16.iop.org/advphys) guides you to extra useful information on the Web.

Advancing Physics AS authors, coordinators and contributors:

Susan Aldridge
Michael Barnett
Richard Boohan
David Brabban
Jim Breithaupt (A-Z)
Philip Britton
James Butler
Peter Campbell
Simon Carson
Simon Collins
Laurence Dickie
Ken Dobson
Diana Emes
Jonathan Foyle
Sarah Grant
Stephen Hall

Stephen Hearn
Lawrence Herklots
David Homer
Neil Hutton
Ian Lawrence
Richard Marshall
John Miller
Jon Ogborn
Steve Pickersgill
Andrew Raw
Laurence Rogers
David Sang
Clare Sansom
Allan Seago
Robert Strawson
Janet Taylor

Project Team:

Philip Britton, Peter Campbell, Simon Carson, Ken Dobson, Ingrid Ebeyer (Administrator), Ian Lawrence, Jon Ogborn (Director), Mary Whitehouse, Catherine Wilson

Production Team:

Penelope Barber, Alan Evans, Huw Johnson, Hayley Liddle, Kevin Lowry, Bridget Pairaudeau, Jamie Symonds, Alison Tovey, Brenda Trigg, Lucy Williams

© **Institute of Physics 2000**

All rights reserved. No part of this publication may be reproduced, stored in a retrieval system, or transmitted in any form or by any means, electronic, mechanical, photocopying, recording or otherwise, without either the prior written permission of the Publishers or a licence permitting restricted copying in the United Kingdom issued by the Copyright Licensing Agency Ltd, 90 Tottenham Court Road, London W1P 0LP.

British Library Cataloguing in Publication Data
A catalogue record for this book is available from the British Library

First published June 2000 by Institute of Physics Publishing, Dirac House, Temple Back, Bristol BS1 6BE, UK
Reprinted 2000, 2001 (twice)

ISBN 0 7503 0585 1

Printed and bound in Spain by Mateu Cromo, Pinto (Madrid)

1 Imaging

Visualising data is an important and rapidly developing aspect of physics. We will give examples of images for:

- use in medicine

- exploring the Universe

- observing the Earth from space

- 'seeing' atoms

The examples will lead you to new ideas about processing and measuring information, and about vision and optical instruments.

Ultimate Reality

A young man said to me:
I am interested in the problem of Reality.

I said: Really!
Then I saw him turn to glance,
surreptitiously,
In the big mirror, at his own fascinating
shadow.

D H Lawrence, in *The Complete Poems* (1994) Penguin

1.1 Seeing invisible things

Ultrasound imaging in pregnancy

The scene is the antenatal department of a hospital. A young woman is there for a checkup. An image of the baby inside her appears on a computer screen. Is everything all right? Seeing a baby develop inside its mother is no longer considered remarkable. Ultrasound scanning is routinely used as a check during pregnancy. Is scanning safe? We have no

● Examples of images in physics

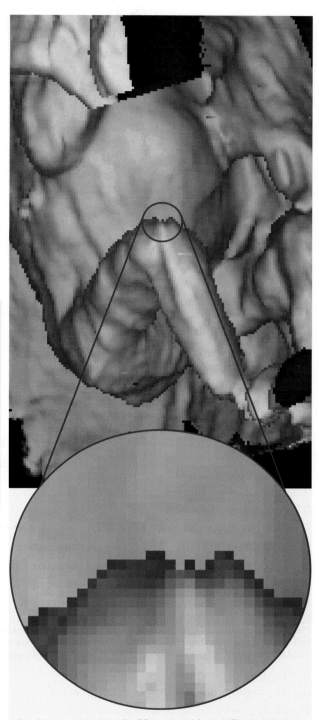

An ultrasound image of a 20 week old foetus. The baby's legs are clearly visible. The head is obscured by part of the wall of the womb. The sound frequency used is several millions of oscillations per second—several MHz—much too high to be audible. The inset enlargement shows the region of the baby's right knee. The discrete square pixels are clearly visible, as are the distinct grey levels.

27	29	33	38	42	45	48
27	28	33	39	168	239	47
27	242	34	235	150	153	218
121	138	213	179	151	146	171
104	83	146	148	124	102	116
32	57	94	113	89	90	113
31	46	62	105	87	86	91
31	44	50	59	76	77	75
31	43	48	54	71	69	68
29	41	46	53	58	61	62

evidence that these scans harm the baby, but having no evidence of an effect does not prove that there is no effect. By contrast, the potentially damaging effects of x-rays are known. So doctors use ultrasound scans rather than x-rays where possible, but still use them with caution, and mothers may choose whether or not to have a scan.

The image looks like a photograph, but it cannot be one; it was made with sound, not with light. You can also see that it is made of small square picture-elements, each a darker or lighter shade of grey. These discrete picture-elements are called pixels. The whole image is made up of 256×256 pixels.

The image was printed from information stored in a computer. This information was worked out from data obtained from the ultrasound scanner. The stored information is just a large array of numbers–one number for each pixel. The number for each pixel decides what shade of grey (or what colour) that pixel will be printed. You can see how darker regions in an image go with larger numbers in the array of numbers.

The array is not a baby, but is *data* about a baby. People are much better at seeing information in images than in sets of numbers, a fact which has revolutionised much scientific work as people increasingly use images to look for important patterns in data. Furthermore, the image is made of numbers, and so it can be modified by manipulating those numbers. The combination of visualisation and numerical processing is enormously effective for extracting and presenting information. It is also changing art and entertainment through computer-generated or computer-manipulated images. So it is worth knowing about how these things are done.

Resolution

The resolution of an image is the smallest size of thing which can be distinguished. The image of the baby has limited resolution. No object smaller than one pixel can be seen. At this stage the baby's limbs will be perhaps 20 mm across. From the image you can see that the width of a pixel represents about 1 mm. The resolution is of the order 1 mm.

The resolution of an image may be very important. Suppose the baby's heart has a slight deformation: the resolution must be good enough to show it up.

The idea of resolution also applies to the grey levels in the scanned image. In this image there are just 256 possible different shades of grey recorded. A change in intensity smaller than this will not be reproduced.

● Improvement in resolution given by the Hubble space telescope

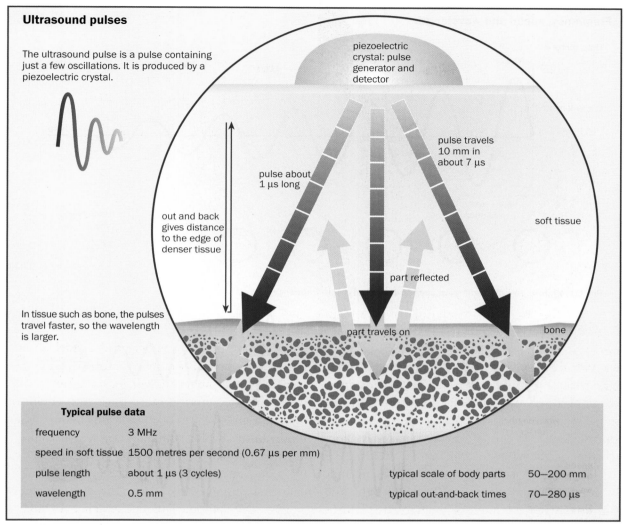

Ultrasound pulses

The ultrasound pulse is a pulse containing just a few oscillations. It is produced by a piezoelectric crystal.

piezoelectric crystal: pulse generator and detector

pulse about 1 µs long

pulse travels 10 mm in about 7 µs

out and back gives distance to the edge of denser tissue

soft tissue

part reflected

In tissue such as bone, the pulses travel faster, so the wavelength is larger.

part travels on

bone

Typical pulse data

frequency	3 MHz		
speed in soft tissue	1500 metres per second (0.67 µs per mm)		
pulse length	about 1 µs (3 cycles)	typical scale of body parts	50—200 mm
wavelength	0.5 mm	typical out-and-back times	70—280 µs

Resolution is a fundamental characteristic of *all* measuring systems. The resolution of any instrument is the smallest difference which is detectable.

Look at a packet of cornflakes or washing powder through a hand lens. See dots of colour? Without the lens your eye does not resolve the dots and you see a smooth colour. Have you seen an Impressionist painting, for example by Seurat? His pictures were composed of dots of primary colour, which seen from a distance merge into a full range of hues. And of course the picture on your television set is also just a dance of coloured dots.

Making a scan with ultrasound

The ultrasound scanner works like a depth-sounder, or like a 'fish-finder' used by trawlers. Short pulses of sound are sent out from the scanner, held in contact with the mother's belly. The pulses are reflected back where the density of the tissue changes. A baby, suspended in fluid in the womb, is quite a good reflector of sound. Bone is a very good reflector. Sound comes back to the scanner from the different parts of the baby, each part producing its own reflection. The further away a part of the baby is, the longer the delay between the pulse being sent and being reflected back.

The 'depths' to be 'sounded' are only about 100 mm, and ultrasound travels at a large speed, taking about 70 µs to go 100 mm in soft tissue. Thus the delay times are small. The pulses must be only about 1 µs long if returning pulses are not to overlap outgoing ones. The time of an oscillation of the wave cannot possibly be longer than the time the pulse exists so the frequency must exceed 1 MHz, that is, one million

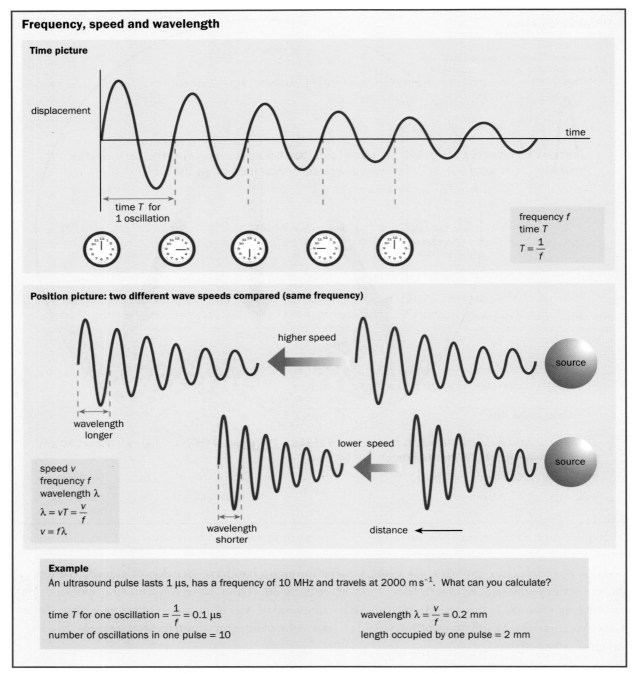

Frequency, speed and wavelength

Time picture

displacement

time

time T for 1 oscillation

frequency f
time T

$$T = \frac{1}{f}$$

Position picture: two different wave speeds compared (same frequency)

higher speed

source

wavelength longer

lower speed

source

speed v
frequency f
wavelength λ

$$\lambda = vT = \frac{v}{f}$$

$$v = f\lambda$$

wavelength shorter

distance

Example

An ultrasound pulse lasts 1 µs, has a frequency of 10 MHz and travels at 2000 m s^{-1}. What can you calculate?

time T for one oscillation $= \dfrac{1}{f} = 0.1$ µs

number of oscillations in one pulse = 10

wavelength $\lambda = \dfrac{v}{f} = 0.2$ mm

length occupied by one pulse = 2 mm

oscillations per second. The highest frequency you can hear is probably about 20 kHz. So ultrasound is ultra-inaudible! The scanning system must be good at measuring very short times.

Ultrasound is generated by rapid electrical oscillations applied to a piezoelectric crystal. When the ultrasound returns and vibrates the crystal, it produces a signal to be picked up in the scanner. You can buy an ultrasonic 'tape measure' at most DIY shops. More about this and other sensors in chapter 2.

The delay times of reflections tell the scanner *where* the denser tissue of the baby is. The strengths of the reflections tell the scanner *how dense* the various tissues are. From the delay times and strengths of reflected pulses a picture of the baby in the womb can be built up.

That's the principle. But it is a big computing job to work out from all those reflections just what is producing them. This is called computer tomography—CT for short—and is widely used in medicine for many kinds of scan.

Try these

1 How many oscillations in 10 seconds from a source of frequency 1 MHz? How many complete cycles in a pulse of ultrasound 3 μs long if the frequency is 2 MHz?

2 If ultrasound takes 70 μs to go 100 mm in soft tissue, how fast does it travel in metres per second? In air, sound travels at about 340 ms^{-1}. How many microseconds does it take to go 100 mm?

3 Suppose one part of a baby is 200 mm in front of another. What is the difference in time between reflections from each part, from a pulse travelling at 70 μs per 100 mm?

4 If the ultrasound pulses are 3.5 μs long, and travel 100 mm in 70 μs, what is the smallest distance one part of a baby can be behind another if the reflections can just be told apart (don't get mixed together in time)?

5 What is the wavelength in soft tissue of ultrasound of frequency 3 MHz travelling at 1500 ms^{-1}? Suggest why detail less than one wavelength in size cannot be seen.

6 The ultrasound image of the baby is a square array of 256 × 256 pixels. If the area imaged is about 250 mm across and high, what is the resolution?

Answers 1. 10 million; 6 **2.** about 1400 ms^{-1}; about 300 μs **3.** 280 μs **4.** about 2.5 mm **5.** 0.5 mm
6. each pixel is about 1 mm by 1 mm

YOU HAVE LEARNED

- How ultrasound scans work, as an example of imaging.

- That computer images are made of pixels, each represented by a number.

- What resolution means.

- How to use wave speeds, frequencies and wavelengths to make estimates and calculations.

wave speeds, ultrasound scanning, frequency, wavelength, pixel

Visible light

UK Schmidt telescope

Infrared

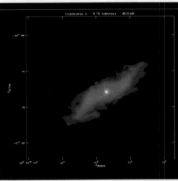

ISOCAM 7 microns

Radio (+ infrared)

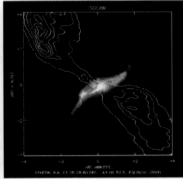

VLA telescope and ISOCAM

Three images of the radio source Centaurus A, one of the strongest radio sources in the sky. It is thought to be a pair of colliding galaxies about 15 million light years away.

Images from the Universe

Until the development of radio astronomy in the 1940s, most news of the Universe was 'read' only through the narrow channel of visible light. Human eyes have evolved to see in just this thin band of the spectrum. Today the Universe is explored at many wavelengths invisible to human eyes. Radio astronomers observe at radio wavelengths; their early radio telescopes were adapted from radar instruments used in wartime. Other telescopes carry instruments to detect infrared, ultraviolet, x-ray and gamma ray emissions. Each new development brings surprises. It is often especially informative to look at the same object at different wavelengths; things invisible at one wavelength can be seen at another.

The image of Centaurus A in visible light can be thought of as a photograph. One galaxy can be seen, but the second which is colliding with it looks like a dusty band across its centre. In infrared, the shape and orientation of this second colliding galaxy can be seen. The data from the radio telescope have been used to plot contour lines of equal radio intensity, showing large lobes of radio activity well outside the regions picked up in visible or infrared radiation. Detailed study of the infrared image gives evidence of a black hole at the very centre, which theory says should be expected to produce the radio emission seen.

How do we know that Centaurus A is 15 million light years away? It was Henrietta Leavitt (1868–1921), working at Harvard, who discovered a way to measure distances to distant galaxies. The brightness of certain stars, called Cepheid variables, varies regularly with time. Leavitt found that the time-period of the variation is related to the brightness of the star. So, given the time-period, the brightness is known. Given the brightness, compared with how bright the star looks, the distance can be found. Henrietta Leavitt headed the Harvard Observatory Department of Photographic Photometry, setting standards for accurate work which were internationally adopted.

The contour lines of intensity of the radio image show large scale structure, at a selected resolution. The resolution of optical and infrared telescopes in space, outside the Earth's atmosphere, is now such

Henrietta Leavitt (left)

Jocelyn Bell Burnell (right)

Regions of the electromagnetic spectrum

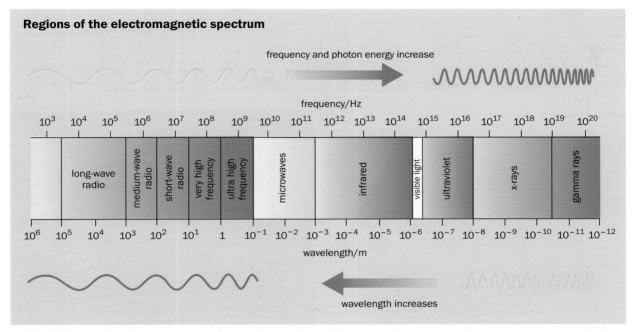

frequency and photon energy increase

frequency/Hz

10^3 10^4 10^5 10^6 10^7 10^8 10^9 10^{10} 10^{11} 10^{12} 10^{13} 10^{14} 10^{15} 10^{16} 10^{17} 10^{18} 10^{19} 10^{20}

long-wave radio | medium-wave radio | short-wave radio | very high frequency | ultra high frequency | microwaves | infrared | visible light | ultraviolet | x-rays | gamma rays

10^6 10^5 10^4 10^3 10^2 10^1 1 10^{-1} 10^{-2} 10^{-3} 10^{-4} 10^{-5} 10^{-6} 10^{-7} 10^{-8} 10^{-9} 10^{-10} 10^{-11} 10^{-12}

wavelength/m

wavelength increases

that there is a chance of detecting the presence of planets circling nearby stars. The question of whether there is life elsewhere in the Universe has become a live issue.

The two astronomers Jocelyn Bell Burnell and Anthony Hewish were joking when they gave the name 'Little Green Men' to regularly pulsing signals they had noticed. Could they be signals from living things elsewhere in the Universe? The thought is tempting but the reality was perhaps more exciting still. They had discovered pulsars—fast spinning relics of stellar explosions, stars made of neutrons packed as densely as particles in an atomic nucleus. The radio telescope they used looks crude and simple: just a large array of aerials joined together. Hammering in a huge number of metal stakes and stringing them with wire was hard work; and when getting on with it they could not anticipate the reward in store, except for hoping that a new channel of information might bring fresh news.

A modern optical telescope uses a light-sensitive microchip instead of a photographic plate. One type is called a 'charge-coupled device' (ccd). It is a screen covered by a million or more tiny silicon picture elements, each of which stores an electric charge when light falls on it. Charge-coupled devices can be

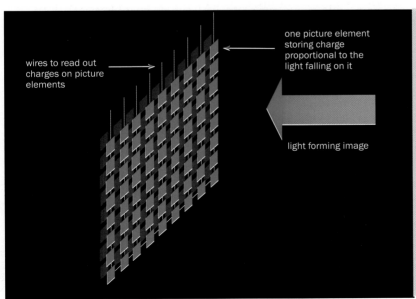

wires to read out charges on picture elements

one picture element storing charge proportional to the light falling on it

light forming image

The structure of a ccd (charge-coupled device). When an image is formed each element stores a charge in proportion to the light which has fallen on it. To record the image, the charges are 'shunted' in sequence from one element to the next, until they reach the edge where the value is read as a potential difference. Thus the whole image is recorded as a sequence of values of potential differences representing charges and so brightnesses. The picture becomes an array of numbers. The technology is continually evolving, and new kinds of digital light-sensitive devices are being developed.

made which are sensitive to wavelengths outside the visible range. The image—a set of values of potential difference, one on each picture element—is 'read out' element by element to be stored as a string of numbers. Back to numbers, again.

Modern digital cameras and camcorders also use ccds or similar devices. Such microsensors (sensors built on a microchip) are becoming essential in modern instrumentation systems (see chapter 2).

Camera on a chip—1998 technology. This chip uses an advance on charge-coupled devices. The hope is to obtain greater sensitivity without an increase in noise. These technologies change all the time and may well have changed again by the time you read these words.

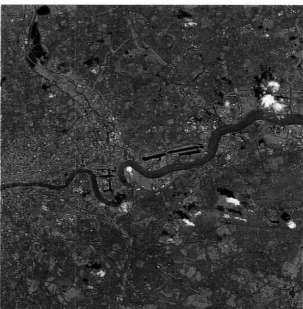

Satellite image of London Docklands taken in infrared. Green vegetation reflects infrared radiation rather well, so areas covered with vegetation show up 'bright' in infrared. In this image, the infrared signal controls the brightness of the red colour. But any colour could have been used. These are called 'false colour' images.

Spy in the sky

International conflict has been an important source of scientific and technological development. During the 'Cold War' (about 1950 to 1990) the nuclear powers tried to detect each other's secret nuclear tests. Satellite imaging played a role in detecting surface traces of underground testing.

School students made front page news tracking the first satellites. Satellite surveying is now big business. CD-ROM maps covering much of the world are widely available.

An infrared satellite can track the growth of vegetation over the seasons, providing valuable ecological information. But it can also detect farmers in Europe growing crops on land subsidised by the European Union to be 'set-aside' (left fallow). To the people who pay the subsidy, this seems like detecting cheating. To the farmer it may feel like spying.

A satellite image of the Earth can easily have a resolution of 10 m or less. This means that you might be able to identify the building where you live, but would probably not be able to detect parked cars. You can estimate the resolution if you 'blow up' the image until the pixels are visible. Then identify some object whose size you can guess, and see how wide it is measured in pixels.

● Satellite images of towns in Europe

Try these

The speed of electromagnetic radiation is 300 000 km per second or $3 \times 10^8 \, ms^{-1}$.
1 MHz is one million (10^6) oscillations per second. 1 GHz is one billion (10^9) oscillations per second.

1   **There are about 31 million seconds in a year. How far does electromagnetic radiation travel in a year?**

2   **Light takes 500 seconds to reach the Earth from the Sun. Light from Centaurus A (page 6) takes 15 million years to reach the Earth. How many times further away is Centaurus A than the Sun? How far away is Centaurus A in kilometres?**

3   **The electromagnetic waves in a microwave oven have a wavelength of about 100 mm. What is the frequency of the generator in a microwave oven? What is the wavelength of electromagnetic radiation with frequency 300 GHz?**

4   **If you could 'freeze' a 3 GHz electromagnetic wave as it travels through space, how many complete waves would there be in a distance of 1 m?**

5   **A 5p coin (18 mm across) held at arm's length (about 1 metre) covers an angle of about one degree. Radio telescopes can distinguish between (resolve) details better than one second of arc apart. One second of arc is 1/60 of a minute of arc, which is 1/60 of a degree. How far away would a 5p coin have to be held to cover one second of arc?**

6   **Centaurus A is 15 million light years away. How big are details which when seen from Earth appear to be 1 second of arc apart?**

Answers 1. about 9×10^{12} km, or 9×10^{15} m which is 1 light-year **2.** about a million million, that is 10^{12}; about 1.4×10^{20} km **3.** 3 GHz; 1 mm **4.** 10 **5.** 3.6 km **6.** about 75 light-years across

YOU HAVE LEARNED

- That astronomical data can be obtained across the entire electromagnetic spectrum.

- How microsensors such as charge-coupled devices are used to convert light into digitally stored images.

- That false colour images can be used to picture invisible radiation.

- More calculations of wavelengths, frequencies and distances.

electromagnetic spectrum, galaxy, pulsar, quasar, radians, resolution, ccd camera

Ernst Mach (1838–1916). 'Mach numbers' are named after him. An aeroplane flying at 'Mach 2.0' is flying at twice the speed of sound in air in those conditions.

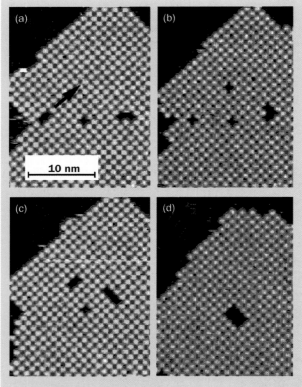

Moving molecules around. A layer of organic molecules has been deposited on a silver surface. There are several places where a molecule is missing from the regular pattern. A scanning tunnelling microscope has been used not only to image the layer of molecules, but also to push them around. Images (a) to (d) show successive stages, as the molecules are moved so as to bring all the empty sites together.

'Seeing' atoms

The Austrian physicist and philosopher Ernst Mach hated what he called 'fictive physics'. Atoms, Mach said, are "things which can never be seen or touched and only exist in our imagination". Challenging those like Ludwig Boltzmann who thought of atoms as real, he would ask, "Haben Sie einen gesehen?"—"Have you seen one?". You should be aware that in Mach's time nobody had a good measurement even of how big atoms are, nor a good theory of what they might be made of. We believe that Mach actually did science a real service, even if in the end he was wrong. *You don't really know,* he said—*it's only an idea.* Science depends on imagination but also on doubt and criticism.

Mach died still disbelieving in atoms just as Max von Laue found the first visual evidence of their existence using x-rays, as Ernest Rutherford with Hans Geiger and Ernest Marsden discovered the nucleus within the atom and as Niels Bohr produced the first 'quantum' theory of atoms.

Nowadays atoms can be 'seen' and 'touched'. They can even be picked up and put down at will. The answer to Mach's question, "Have you seen one?" can be, "In a way, yes".

In chapter 5 you will learn about what can be understood from imaging atoms. For now, here is a set of images showing molecules not just being seen but also being deliberately moved about.

There are several kinds of scanning microscope able to make images on the atomic scale, or near to it. One is the scanning tunnelling microscope. This has an ultrasharp needle point which is scanned in zigzags across the surface of a material. When the tip is close to the atoms of the surface, electrons can 'tunnel' across the gap so that there is an electric current between surface and tip. The tip is moved up or down so as to keep the tunnelling current constant. A record of the movements of the tip is then a record of ups and downs at the atomic scale on the surface.

● Examples of images in physics

atomic force microscope, scanning electron microscope, scanning tunnelling microscope

Scanning tunnelling microscope

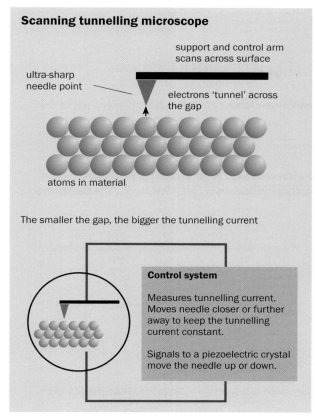

support and control arm
scans across surface

ultra-sharp
needle point

electrons 'tunnel' across
the gap

atoms in material

The smaller the gap, the bigger the tunnelling current

Control system

Measures tunnelling current.
Moves needle closer or further
away to keep the tunnelling
current constant.

Signals to a piezoelectric crystal
move the needle up or down.

Here you see a part of the sky with many galaxies, plotted so that if they are distributed evenly in space around us, they will be plotted evenly over the diagram. There is still debate about what these data say about the structure of the Universe. Are the galaxies randomly distributed? Do they form structures like bubbles with voids in between? Data images like these are also images for the mind's eye.

Does a scanning tunnelling microscope 'see' atoms? What is actually observed is the electron current, not the atoms.

Images for the mind's eye

If 'seeing atoms' seems unlikely, then 'seeing a theory' must sound even more peculiar. It isn't. Today, very often theories of things never seen, such as the way 'black holes' might evolve, galaxies might form, or DNA molecules might fold up, are best pictured visually. A computer crunches numbers to simulate what might happen, and then uses the numbers to make an image. The image is not a picture of the result: the image *is* the result.

Computer-generated image of a tornado developing. This storm never blew off a roof. It is not a picture of a cloud. It is not an image of data about a cloud. It was made entirely by calculation, using a computer model of the development of storms.

Try these

An atom is of the order of 0.1 nanometre across (1 nm = 10^{-9} m).

1   How many atoms are there in (a) the 20 mm width of a postage stamp, (b) the 0.1 mm width of a pencil line?

2   What must be the resolution of a scanning tunnelling microscope used to obtain the image of molecules on page 10? (Look at the scale, and remember that one of these molecules is only a few tenths of a nanometre across.)

3   Microchips can be fabricated with elements on a scale of 1 μm. How many atoms across is such an element?

4   If atoms were as big as apples, how tall would you be?

5   Going up a multiple of 10 in size for each step, how many steps is it from an atom to your little fingernail?

6   Make a list of images you have seen (on TV, in magazines, books, on the Web) which are visual presentations of data.

Answers 1. (a) about 200 million; (b) one million **2.** around 0.1 nm **3.** ten thousand **4.** around a million kilometres **5.** say eight **6.** for example, an animated weather map on TV

YOU HAVE LEARNED

- How a scanning tunnelling microscope works, as an example of imaging on the very small scale.

- To make calculations and estimates using units such as nanometres.

- That images of atoms are further examples of images constructed from data.

- That images are made of theories as well as of things.

atom, scanning tunnelling microscope

1.2 Information in images

Numbers in computers are stored as on- or off-values. 'On' may be a high potential difference, 'off' may be low. This is a *binary* system. The two values may be thought of as two digits 1 and 0.

During the 1940s and 1950s, when computers were first coming into use, Claude Shannon, working at Bell Telephone Laboratories, devised a theory of how to measure information. One of his ideas was that the amount of information can be measured by the amount of storage it needs. If one pixel was simply 'bright' or 'dark', only one memory location storing a 1 or a 0 would be needed. John Tukey named this one 'bit' of information (contracting 'binary digit' to 'bit'). But often a pixel can have, say, 256 different shades of grey. One of 256 alternatives can be stored as a number from 0 to 255. That needs more memory space for each pixel. In fact it needs just eight memory locations—eight bits.

Claude Shannon.

Shannon's mathematical theory of communication has proved important in linguistics and cryptography, as well as in the design of communication systems.

It is a curious fact that Shannon's theory measures information in a message without any regard to the 'meaning' or content of the message. This enables the amount of information in a picture to be compared with that in a text, a coded message or a table of numbers.

Bytes

One bit of information is one choice: 0 or 1. Eight bits, giving 256 possibilities, is a useful slice of memory. It is enough to give a code number to every key on the keyboard, including capitals and lower case, numbers and punctuation marks. Early in the history of computing it was chosen as a unit of its own—one **byte**. Thus a 10 kbyte word-processed file of yours will contain around ten

thousand characters—and so about 1600 words (English words average about six characters each).

Actually, in computing, one kilobyte is 1024 bytes, not 1000 bytes, because 1024 is a simple power of two. Similarly one megabyte is a little more than 10^6 bytes.

Amount of information

The better the resolution of an image, the more information it contains. A digital colour camera image can have over a million pixels, each using 3 bytes (24 bits) for colour information about each pixel. So the image may need 3 megabytes of storage. Storing images on a computer very rapidly uses up the storage capacity, even if it has 1 gigabyte or more. And processing images places big demands on memory while the image is being worked on, because many operations have to be performed.

Computers claim to offer 'millions of colours'. How few bits are needed to represent so many colours? The number n of bits must be big enough

---

Number of alternatives = 2$^{\text{number of bits}}$

The box 'Bits and bytes' (*overleaf*) shows that the number of alternative values which can be represented grows rapidly as the amount of memory used increases.

If 8 bits of information are used:

number of alternatives = 2^8 = 256

In general, if the amount of information is I bits:

number of alternatives = 2^I

If the number of bits *increases by one*, the number of alternatives *doubles*. Information is measured on a 'plus' scale; number of alternatives on a 'times' scale:

'Plus' scale (linear)	'Times' scale (logarithmic)
amount of information	number of alternatives
increase by equal additions	increase in equal multiples
amount = I	number of alternatives = 2^I
amount = $\log_2 N$	number of alternatives = N

$\log_2$ **(number of alternatives) = information I = number of bits**

---

that 2^n is more than a million. Try values on your calculator. The number n needed turns out to be only 20. So if each pixel has to store one of at least a million possibilities, between two and three bytes are needed for each (2 bytes is 16 bits; 3 bytes is 24 bits). The number 20 is approximately the *logarithm* to the base 2 of one million.

You get information from a message when it tells you one thing rather than another. It is informative because it communicates one possible alternative out of many.

Bits and bytes
One byte stores 256 alternatives

8 bits = 1 byte				4 bits		2 bits	1 bit	Decimal value	Number of alternatives
0	0	0	0	0	0	0	0	0	
0	0	0	0	0	0	0	1	1	
0	0	0	0	0	0	1	0	2	$2^1 = 2$
0	0	0	0	0	0	1	1	3	
0	0	0	0	0	1	0	0	4	$2^2 = 4$
0	0	0	0	0	1	0	1	5	
0	0	0	0	0	1	1	0	6	
0	0	0	0	0	1	1	1	7	
0	0	0	0	1	0	0	0	8	$2^3 = 8$
. . .								. . .	
0	0	0	0	1	1	1	1	15	
0	0	0	1	0	0	0	0	16	$2^4 = 16$
. . .								. . .	
0	0	0	1	1	1	1	1	31	
0	0	1	0	0	0	0	0	32	$2^5 = 32$
. . .								. . .	
0	0	1	1	1	1	1	1	63	
0	1	0	0	0	0	0	0	64	$2^6 = 64$
. . .								. . .	
0	1	1	1	1	1	1	1	127	
1	0	0	0	0	0	0	0	128	$2^7 = 128$
. . .								. . .	
1	1	1	1	1	1	1	1	255	
1	0	0	0	0	0	0	0	256	$2^8 = 256$

Information engineers use logarithms to measure amounts of information. Sound engineers use logarithms to measure sound intensity in decibels. Astronomers use logarithms to measure brightness of stars in magnitudes. They are often a good way to make plots of quantities which span many orders of magnitude. You will use logarithms a lot. They make some kinds of thinking much easier, once you have got your mind round them.

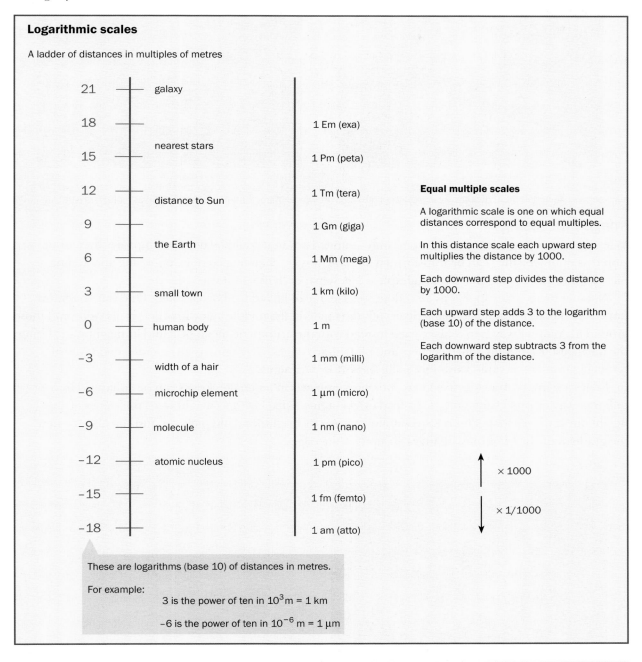

Logarithmic scales

A ladder of distances in multiples of metres

Log	Label	Unit
21	galaxy	
18		1 Em (exa)
15	nearest stars	1 Pm (peta)
12	distance to Sun	1 Tm (tera)
9		1 Gm (giga)
6	the Earth	1 Mm (mega)
3	small town	1 km (kilo)
0	human body	1 m
-3	width of a hair	1 mm (milli)
-6	microchip element	1 µm (micro)
-9	molecule	1 nm (nano)
-12	atomic nucleus	1 pm (pico)
-15		1 fm (femto)
-18		1 am (atto)

Equal multiple scales

A logarithmic scale is one on which equal distances correspond to equal multiples.

In this distance scale each upward step multiplies the distance by 1000.

Each downward step divides the distance by 1000.

Each upward step adds 3 to the logarithm (base 10) of the distance.

Each downward step subtracts 3 from the logarithm of the distance.

× 1000

× 1/1000

These are logarithms (base 10) of distances in metres.

For example:

3 is the power of ten in 10^3 m = 1 km

−6 is the power of ten in 10^{-6} m = 1 µm

Logarithms, a way to use small numbers to think about big numbers by turning multiplying into adding, are a clever and elegant idea. They were invented long ago by John Napier (1550–1617), an independent-minded Scot fascinated by the idea of doing all calculations by machine. His logarithms made the tedious astronomical calculations of the time much easier—the only calculators then were people. Napier believed no less strongly in astrology and divination, was fiercely involved in the religious disputes of the time, and designed a primitive tank he hoped would be used in the war against Philip II of Spain.

Comparison of images of stars in a cluster seen from a ground-based telescope, and by the Hubble space telescope. The sharpness and number of stars detectable are both much improved. Also, a larger number of faint stars can be detected.

Image processing

When images were mainly photographs, only a limited amount could be done to 'improve' them. But stored as numbers, they can be transformed in many ways—changing colours, removing and adding pieces. You can't be sure that a photograph in a newspaper 'tells it as it was'.

Noise in images can appear as a random speckle. It can be reduced by *smoothing*. One way to remove noise is to replace the value of each pixel with the median (the middle value in order) of its value and those around it. Another way is to replace each instead with the arithmetic mean of it and its neighbours. Using the mean also rounds off sharp corners and blurs sharp edges. Averaging is in general a good way of removing random or rapid variations in all sorts of experiments.

Edges are important clues if you are looking for objects in an image. Edges can be enhanced by a different way of processing images. Instead of averaging a pixel with its neighbours, the values of the neighbours are *subtracted*. This removes uniform areas of brightness, and picks out just the places where the gradient of the brightness changes abruptly—the edges.

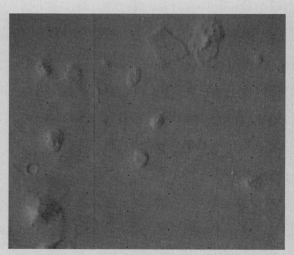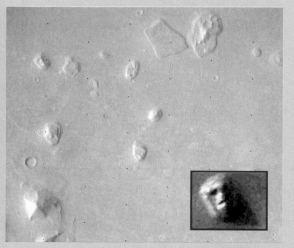

The left hand image is an unprocessed image of the surface of Mars, taken by the Viking 1 Orbiter. On the right is the same image with noise removed, and contrast enhanced. There has been controversy over the 'face-like' markings in one part of the image. Are they real, or are they introduced by the image processing?

Image processing

Smoothing sharp edges

rule

replace each pixel by the mean of
its value and those of its neighbours

effect if there is an edge

Removing noise

rule

replace each pixel by the median of
its value and those of its neighbours

effect if there is noise

Finding edges

rule

	−1	
−1	**+4**	−1
	−1	

subtract the N, S, E and W neighbours
from 4 times the value of each pixel.

effect if there is an edge

This is called the Laplace rule. It
detects changes in the gradient of
brightness.

effect if there is no edge

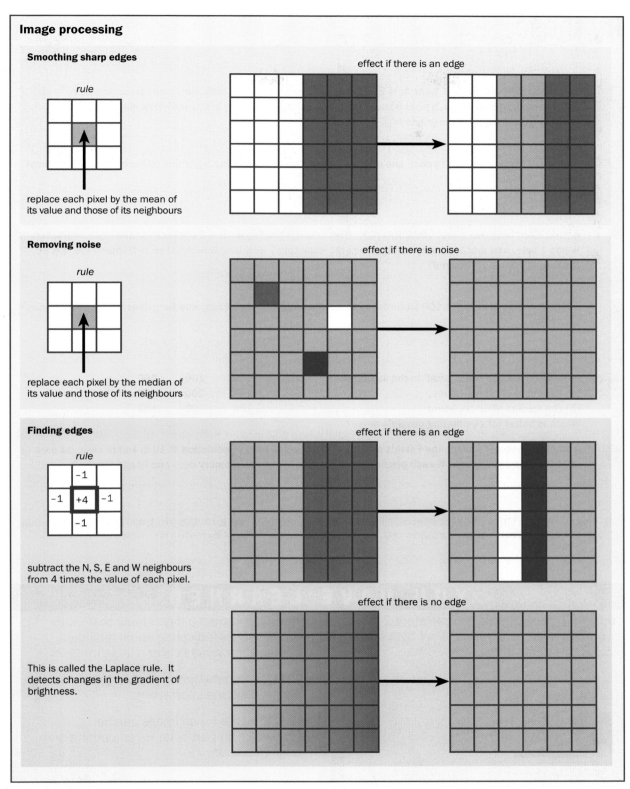

- Spreadsheet models: Image processing
- Image processing: The surface of Mercury

- Astronomical images: Problems of noise and
resolution

Try these

1   The ultrasound scan image, page 1, is 256 pixels wide and 256 pixels high. How many pixels are there in all? If the number stored for each pixel occupies 1 byte, how much memory is needed? How many such images could be stored on a 1.4 megabyte high density disc?

2   How many bits are needed to code one of 1024 alternatives? What is the logarithm to base two of the decimal number 1024?

3   A good sized book has about 100 000 words. If an average English word is 6 characters, and each character needs 1 byte, how much memory storage does the book take? How long would it take to transmit the text at the rate of 1 Mbit per second?

4   Suppose the book also has 100 pictures, each needing 1 Mbyte of storage. How long does the transmission take now?

5   In this array, replace the 'pixel' in the middle by
(a) the mean of all the values,
(b) the median of all the values.
Which is better for eliminating possible noise?

100	100	100
100	200	100
100	100	100

6   A satellite system to image the Earth's surface is designed to have a resolution of 10 m and to cover an area of 100 km^2 in each image. If each pixel requires 3 bytes, how much memory does one image require?

Answers 1. 65 536 pixels; about 65 kbyte; about 20 such images can be stored **2.** 10; 10 again **3.** 600 kbyte; nearly 5 seconds **4.** over 800 seconds, or nearly a quarter of an hour **5.** (a) 111; (b) 100; the use of the median eliminates a 'stray' value completely **6.** 3 Mbyte

YOU HAVE LEARNED

- That I bits of information can store 2^I different values; that 8 bits = 1 byte stores 256 different values.

amount of information, bit, byte, logarithm, logarithmic plot, median, mean, noise

- How to construct a logarithmic scale with equal spaces indicating equal multiples, useful when values cover a large range.

- That logarithms turn multiplying and dividing into adding and subtracting.

- That the data for an image can be processed, in particular for smoothing and edge detection.

1.3 With your own eyes

Images and visualisation are important because eyes, and the brains of which they are an extension, are such clever instruments.

The eye is like a camcorder. But it is like a clever camcorder with intelligent computerised control of direction, focus and resolution, and with on-board chips processing signals before sending them on to the computer.

You can learn several surprising things about your intelligent eye by some simple tests.

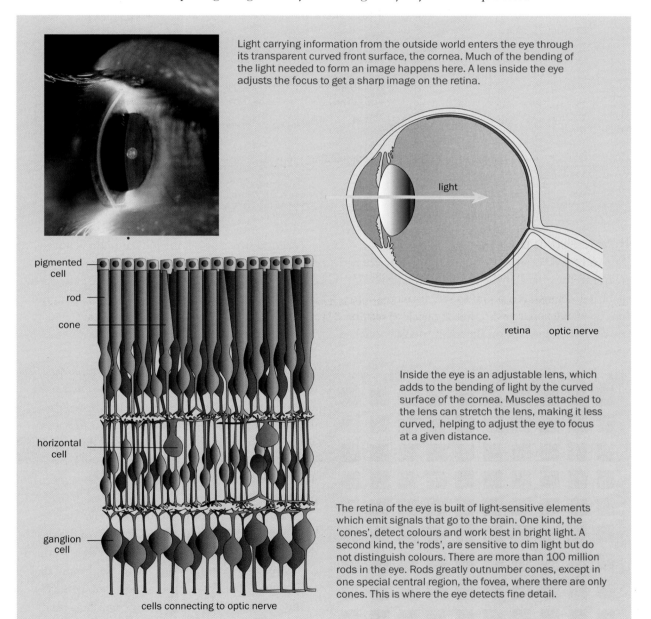

Light carrying information from the outside world enters the eye through its transparent curved front surface, the cornea. Much of the bending of the light needed to form an image happens here. A lens inside the eye adjusts the focus to get a sharp image on the retina.

light

retina   optic nerve

pigmented cell
rod
cone
horizontal cell
ganglion cell

cells connecting to optic nerve

Inside the eye is an adjustable lens, which adds to the bending of light by the curved surface of the cornea. Muscles attached to the lens can stretch the lens, making it less curved, helping to adjust the eye to focus at a given distance.

The retina of the eye is built of light-sensitive elements which emit signals that go to the brain. One kind, the 'cones', detect colours and work best in bright light. A second kind, the 'rods', are sensitive to dim light but do not distinguish colours. There are more than 100 million rods in the eye. Rods greatly outnumber cones, except in one special central region, the fovea, where there are only cones. This is where the eye detects fine detail.

● A collection of visual illusions

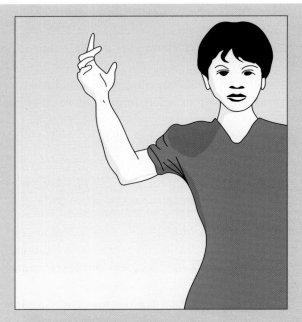

Look straight ahead and hold your hand out to one side, but don't look at it. How many fingers can you count? You probably can't count them.

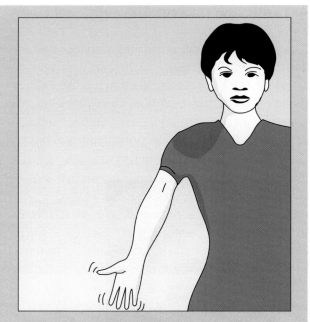

Look straight ahead and hold your hand by your side. You probably don't see it. If you waggle it you should see the movement. But you do not see the hand. The brain just detects 'something moving'!

How to see in the dark

The rods and cones are less and less tightly packed nearer to the edges of the retina, so the resolution of off-centre vision is poor. To see distinctly, you have to move your eyes so that the desired part of the retinal image falls on the fovea, densely packed with cones.

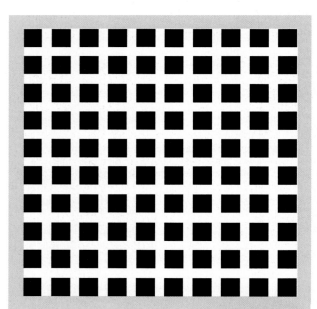

This grid of black squares looks as if it has greyish patches at the crossing points. Artists such as Bridget Riley have exploited optical illusions in their work.

A good way to see faint stars at night is not to look directly at them. If you look just to one side, the star will look brighter. This is because the rods in the retina are sensitive to low levels of light, but there are none of these rods in the fovea.

Your retina 'thinks'

The rods and cones in the retina are connected to those near them. The connections are *inhibitory*; a lot of light on one rod or cone turns down the sensitivity of its neighbours. The rods and cones 'seeing' the crossing points of the grid have their sensitivity turned down because many of those near them are also brightly lit. So you 'see' greyish patches at those places.

This system of connections works like the 'sharpening' of an image (see page 16). It makes the eye into a good detector of sharp edges. Indeed the eye exaggerates the contrast of bright and dark at an edge. This clever processing—done entirely in the retina—helps you to see the world as made out of well-defined objects with definite boundaries. The grey patches which you 'see' in the grid are your retinal edge-detectors working overtime.

Other effects like this have been exploited by the painter Bridget Riley. Some of her still paintings look as if the pattern is moving.

Shaping light to make images

Optics must have started in ancient times when people found glass fused from sand in their fires. The first telescopes, in the hands of Galileo Galilei (1564–1642), helped shake the foundations of the then current view of the Universe. Some refused to believe that Galileo could see sunspots, the phases of Venus and the moons of Jupiter. Nor did they like the idea that the Earth might be just one body in the heavens amongst others. Disturbing thoughts arose: might not other heavenly bodies have life on them?—a question not answered even today. And worse, what religion would a God have given such living beings? The Italian monk Giordano Bruno was burnt at the stake for this heretical idea.

Images are made using lenses or mirrors. There are several ways of thinking about how lenses or mirrors shape light to make images. We have decided to use mainly the wave point of view, but will link it to the ray point of view which you will find in many other books. There is value in more than one point of view:

- one explanation helps with some problems; another helps more with others.
- there really are several different ways of thinking about light (see chapters 6 and 7).

In the ray point of view, light is tracked along rays which go straight, until the light is reflected at a surface or is bent as it passes from one substance into another. Rays from a point source spread out

Anton van Leeuwenhoek (1632–1723), working in Delft in Holland, saw with his single-lensed microscopes, which could magnify up to 300 times, a whole new world of the very small which has by now expanded into microbiology and biotechnology. The microscope in the photograph is of a kind developed from those made by van Leeuwenhoek.

in all directions, getting further and further apart. From the wave point of view, waves spread out in all directions from a point source like the circular ripples from a stone thrown in a pond.

The connection between the two points of view is that a ray is a line pointing along the direction of motion of the wave front. Rays are always at right angles to the wave fronts at the point where they cross. Rays are convenient construction lines, not real light paths, though the beam from a laser does stay quite close to a single straight line.

What a lens does

A lens used as a 'burning glass' brings light across the whole surface of the lens down into a tiny image of the Sun, at the point called the focus (named after the Latin word for 'hearth'). In ray-language, the burning glass works by bending rays, bringing

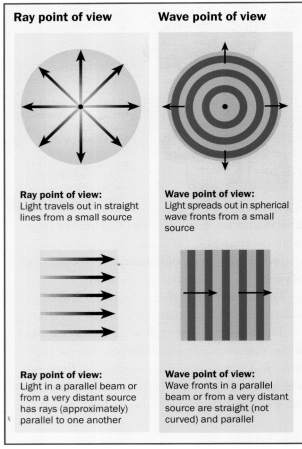

Ray point of view

Ray point of view:
Light travels out in straight lines from a small source

Ray point of view:
Light in a parallel beam or from a very distant source has rays (approximately) parallel to one another

Wave point of view

Wave point of view:
Light spreads out in spherical wave fronts from a small source

Wave point of view:
Wave fronts in a parallel beam or from a very distant source are straight (not curved) and parallel

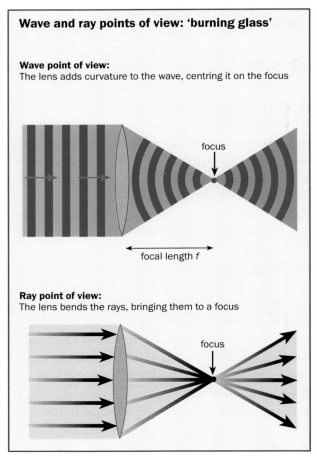

Wave and ray points of view: 'burning glass'

Wave point of view:
The lens adds curvature to the wave, centring it on the focus

focus

focal length f

Ray point of view:
The lens bends the rays, bringing them to a focus

focus

them from being parallel to coming together at a point. In wave language, the lens works by changing the curvature of the wave, changing it into spherical 'ripples' converging on the focus.

The lens makes the waves more curved by slowing down the middle more than the edges. Waves going slowly through the thick glass in the middle of the lens get delayed more than waves going through the thinner glass near the edges of the lens. Waves through the middle get 'left behind' a little. More about this in chapters 4 and 5.

The distance from lens to focus is called the *focal length f*. A powerful lens has a small focal length. A typical camera lens has a focal length of about 50 mm. A spectacle lens for a long-sighted person may have a focal length of 250 mm or so.

A lens changes how curved the waves are. If they are not curved at all when they get to the lens, they become spheres centred on the focal point after going through the lens. The radius of the spherical wave fronts just after getting through the lens is f, the focal length. The curvature of a sphere of radius r is $1/r$, bigger when r is smaller. So the lens has added curvature $1/f$ to the wave fronts. The *power* of a lens, measured in dioptres, is $1/f$, i.e. the curvature it adds, with f measured in metres.

Thus a camera lens with focal length 50 mm has a power of $1/(50 \times 10^{-3}) = 20$ dioptre. A spectacle lens with focal length 0.5 m has a power of $1/0.5 = 2$ dioptre. There are many other examples in physics of such pairs of reciprocal quantities, where $1/a = b$ and $1/b = a$ (see resistance and conductance in chapter 2).

Photographers and astronomers mostly work with focal lengths. Opticians work with lens powers in dioptres. You take your choice, but it is handy to be able to speak both languages. The big advantage of lens powers is that they add up: the power of two thin lenses put together in contact is the sum of the powers of each alone.

How a lens makes 'a little picture'

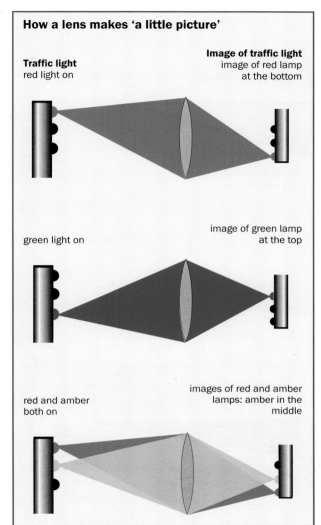

Traffic light
red light on

Image of traffic light
image of red lamp
at the bottom

green light on

image of green lamp
at the top

red and amber
both on

images of red and amber
lamps: amber in the
middle

The making of images

A lens makes an image 'like a little picture' because light goes through the lens from every part of the object in front of the lens. Light from a particular point of the object comes together again behind the lens at a particular point in the image. Because the general direction of travel of the light is not altered by the lens, but only the curvature of the waves, points high up on the object come low down on the image, and points on the right of the object come to the left of the image.

To see detail, the important thing is the size of the image on your retina. Bring your eye quickly closer to this page. The objects on the page 'zoom' up in size as the image on the back of your eye gets bigger. But they get fuzzy too—your eye has not enough power to add the needed curvature to the waves to bring them to a focus on the retina. The

answer is more power: add a magnifying glass in front of the eye. Then you can have the image on your retina big and in focus.

You can't get closer to a galaxy to see it better. So a telescope uses a lens or mirror to make an image of the star close to your eye. You then look at this image with a magnifying glass (the eyepiece).

The rule for how lenses shape light

The rule for how a lens shapes light is simple:

A lens changes the curvature of the wave fronts by a fixed amount, $1/f$,
or

curvature of waves going out
= curvature of waves coming in
+ curvature added by lens

This can be translated into a rule for distances of image and object from the lens, and the focal length, which is more natural from the ray point of view. The rule is:

$$\frac{1}{v} = \frac{1}{u} + \frac{1}{f}$$

where u is the distance of the source from the lens, v is the distance of the image of the source from the lens, and f is the focal length.

Read as a formula linking reciprocals of distances, the lens rule looks hard to understand. Read as a rule which says that a lens changes the curvature of wave fronts by a fixed amount, it makes better sense.

The rule for the signs of lens distances is like the rule for signs when drawing graphs:
1. Measure from the lens
2. Count distances to the right positive
3. Count distances to the left negative
Because they are like Cartesian graph rules, these rules are called the Cartesian sign convention.

If the source is far away, the wave fronts are flat and the lens makes them curved. Bring the source up to the focal point of the lens, and the lens adds curvature to the negatively curved wave fronts from the source, making them flat. In between, both incoming and outgoing wave fronts are curved, and the difference between their curvatures is the amount $1/f$ added by the lens.

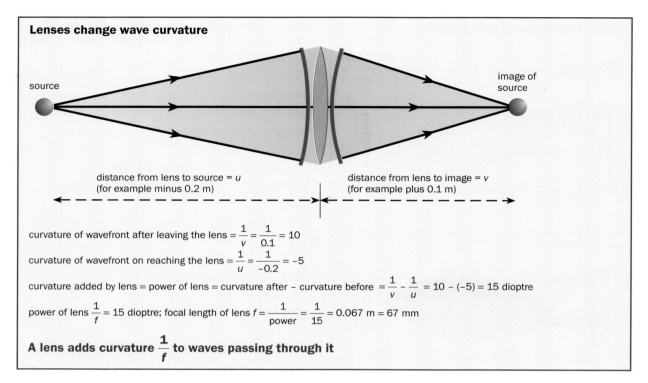

Lenses change wave curvature

source

image of source

distance from lens to source = u
(for example minus 0.2 m)

distance from lens to image = v
(for example plus 0.1 m)

curvature of wavefront after leaving the lens $= \dfrac{1}{v} = \dfrac{1}{0.1} = 10$

curvature of wavefront on reaching the lens $= \dfrac{1}{u} = \dfrac{1}{-0.2} = -5$

curvature added by lens = power of lens = curvature after − curvature before $= \dfrac{1}{v} - \dfrac{1}{u} = 10 - (-5) = 15$ dioptre

power of lens $\dfrac{1}{f} = 15$ dioptre; focal length of lens $f = \dfrac{1}{power} = \dfrac{1}{15} = 0.067$ m = 67 mm

A lens adds curvature $\dfrac{1}{f}$ to waves passing through it

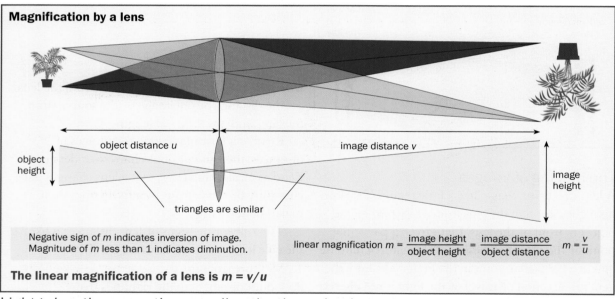

Magnification by a lens

object height

object distance u

image distance v

triangles are similar

image height

Negative sign of m indicates inversion of image.
Magnitude of m less than 1 indicates diminution.

linear magnification $m = \dfrac{\text{image height}}{\text{object height}} = \dfrac{\text{image distance}}{\text{object distance}}$ $m = \dfrac{v}{u}$

The linear magnification of a lens is $m = v/u$

Light takes the same time on all paths through a lens

There is another very remarkable way of looking at what light does as it goes through a lens. It is this: light clocks up exactly the same time on all paths from a point on a source to the corresponding point on the image. How come? The angled outer paths are obviously longer than ones through the centre. So why don't they take longer? The answer is that through the centre the light travels through more glass, going where the lens is fattest, and because it goes more slowly in glass it takes longer to travel in the glass. At the edges of the lens, the light travels further, but in air where it goes quickly. The two effects balance out.

It has to be like this. After all, the lens just adds curvature to the wave, and manages to bring the whole wave 'together' at the image, with all parts of the wave front still in step. Keeping the wave front 'together' *is* making all parts of it take the same time to travel. Chapter 7 shows how this idea can be taken further still, into quantum behaviour.

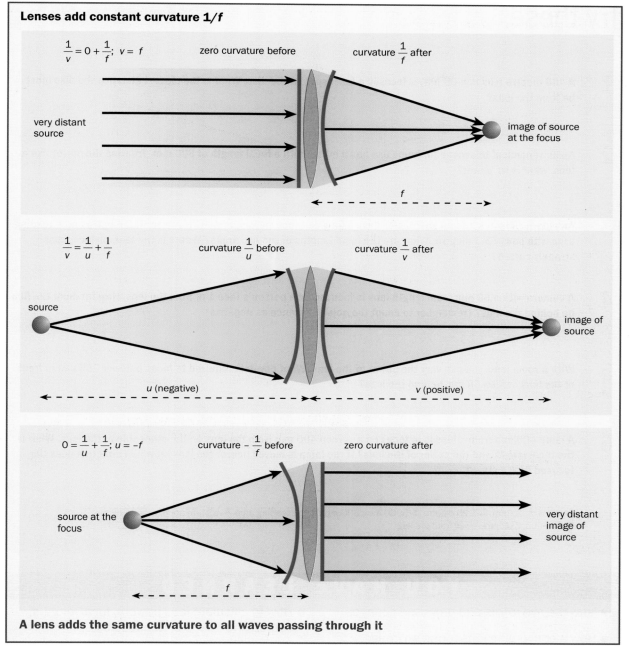

Lenses add constant curvature 1/f

$\dfrac{1}{v} = 0 + \dfrac{1}{f};\ v = f$ zero curvature before

curvature $\dfrac{1}{f}$ after

very distant source

image of source at the focus

f

$\dfrac{1}{v} = \dfrac{1}{u} + \dfrac{1}{f}$ curvature $\dfrac{1}{u}$ before

curvature $\dfrac{1}{v}$ after

source

image of source

u (negative)

v (positive)

$0 = \dfrac{1}{u} + \dfrac{1}{f};\ u = -f$ curvature $-\dfrac{1}{f}$ before

zero curvature after

source at the focus

very distant image of source

f

A lens adds the same curvature to all waves passing through it

Keeping things simple

There is much more to the story of how the eye works than we have told here. The theory of lenses provided here is also very simplified. It works only if the lens is thin, if the width of the lens is small compared to its focal length, and if rays pass at small angles to the axis of the lens. For more complicated lenses, such as the zoom lens on a camera, professionals use computer ray-tracing, or treat the lens as 'transforming' source into image. Computer ray-tracing is also used to generate realistic-looking artificial scenes, with shadows and highlights all in the right places.

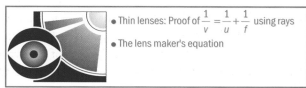

- Thin lenses: Proof of $\dfrac{1}{v} = \dfrac{1}{u} + \dfrac{1}{f}$ using rays
- The lens maker's equation

Try these

1   A 100 dioptre lens in a CD player focuses a laser spot on the disc. What is the closest distance the disc must be from the lens?

2   An astronomical telescope for home use has a mirror with a focal length of 800 mm. Treating the mirror like a lens, what is its power?

3   An elderly person's bifocal spectacles have a long-distance viewing area with power −0.2 dioptre and a reading area with power 3.3 dioptre. What are the focal lengths of the two areas? Where is the lens surface most strongly curved?

4   A camera with a 50 mm focal length lens is focused on a person's face 1 m from the lens. How far must the film be behind the lens? (remember to count the object distance as negative)

5   With a zoom lens, you can vary the power of the lens. What power is required to focus a flower 250 mm in front of the lens on film 50 mm behind the lens?

6   A lamp 400 mm from a lens is in focus on a screen 400 mm from the lens on the other side of the lens. What is the focal length and the power of the lens? If the lamp is moved nearer the lens, in which direction does the focused image move?

Answers 1. 10 mm **2.** 1.25 dioptre **3.** 5000 mm; 300 mm; in the reading area **4.** 53 mm approx. **5.** 24 dioptre **6.** 200 mm, 5 dioptre, away from the lens

YOU HAVE LEARNED

- How the eye forms and processes images.
- How a converging lens forms an image.

- How to calculate lens powers, focal lengths, source and image distances and magnification.

eye, lens, refraction, refractive index, focal length, power of lens, linear magnification

Summary checkup

✓ Waves

- A wave of frequency f travelling at speed v has wavelength λ given by $v = f\lambda$.

✓ Images and information

- Images can be formed with many kinds of signals, including ultrasound and all parts of the electromagnetic spectrum.

- Images can be recorded electronically by microsensors; an example is the charge-coupled device (ccd).

- Images on the atomic scale can be recorded by scanning methods; an example is the scanning tunnelling microscope.

- Images can be stored as an array of pixels, each defined by a number.

- Images can be smoothed by suitable averaging; images can also be sharpened by locating edges.

✓ Logarithms

- Quantities which cover a large range of values can usefully be displayed on a logarithmic ('times') scale. Prefixes for scientific units (e.g. micro-, milli-, kilo-, mega-) are chosen at multiples of 1000.

- 1 bit of information is 2 choices (0 or 1); 1 byte of information contains 8 bits (256 alternatives); amount of information I provides $N = 2^I$ alternatives.

✓ Eye and lenses

- The eye is like an intelligent video camera, sending out a stream of processed signals. It detects edges and movement.

- A converging lens adds a constant curvature to light falling on it. The curvature added is the power of the lens. $\text{Power of lens} = \dfrac{1}{f}$

- Image and object distances $\dfrac{1}{v} = \dfrac{1}{u} + \dfrac{1}{f}$ (Cartesian sign convention).

- The linear magnification of a lens $m = \dfrac{v}{u}$

- Light takes the same time to travel on all paths from a point on the source to the image via the lens (or mirror).

Questions

1 **A digital camera is advertised as having a lens of focal length 5 mm and as producing an image of 1280 × 960 pixels. It is a 'wide angle' camera, and a human face 150 mm from the lens fills the picture area.**

   (a) What is the power of the lens, in dioptre?

   (b) Explain why when the face is in focus the light-sensitive surface will be only a little more than 5 mm from the lens.

   (c) What is the number of pixels in the image?

   (d) Estimate the dimensions of the picture area of the light-sensitive chip in the camera.

   (e) Estimate the dimensions of one picture element in the light-sensitive chip.

   (f) Estimate the scale of features in the face which can be resolved in the picture. Could a person's eyelashes be resolved?

   (g) If the camera stores three colours for each pixel, each with 256 levels of intensity, how big will a picture file be?

2 **Make a logarithmic ('times') scale of the size (i.e. linear dimensions) of living organisms from viruses to whales. Make each point on the scale ten times larger than the one before.**

   (a) Try to place a representative organism at each point on the scale.

   (b) Suggest reasons why there are no organisms ten times smaller than viruses.

   (c) Suggest a reason why there are no organisms ten times larger than whales.

   (d) Show how you can easily add scales for the volume and mass of organisms (you may suppose their density is close to that of water, 1000 kg m^{-3}).

3 **Spectacles for people with long sight use converging lenses.**

   (a) Draw a diagram to show how wearing these spectacles helps such a person to focus sharply on the print in a book close to them.

   (b) Describe how to measure approximately the power of the lenses in such spectacles.

   (c) If the eye has a length of about 25 mm from front to back, estimate roughly the power of the combination of cornea and eye lens.

   (d) A combination of two thin lenses in contact has a power approximately equal to the sum of the powers of the two lenses. Suggest an explanation of this rule. Translate it into an expression for the focal length F of the combination in terms of the focal lengths f_1 and f_2 of the two lenses.

4 **Give your own example of a scientific imaging system (not necessarily using light) producing a computer-storable image.**

   (a) Explain how it works, and how the image is produced.

   (b) What limits the resolution of the image (the detail which can be seen in it)?

   (c) Suggest one advantage and one disadvantage of the system over others.

5 **A balloon filled with carbon dioxide, in which sound travels more slowly than in air, can be used to focus sound waves, like a lens used to focus light.**

   (a) Sketch a diagram of sound waves from a distant source passing through the balloon, showing how they come to a focus.

   (b) Explain how the effect depends on sound travelling more slowly in carbon dioxide than in air.

6 **This is an image of atoms forming a grain boundary. The image is spoiled by a good deal of random noise, which shows as a speckled appearance.**

   (a) What approximately must be the resolution of the image?

   (b) How could you reduce the noise in the image? In doing so, how would new values for the intensity in each pixel be calculated?

   (c) Suggest one other process which could be used to enhance the image.

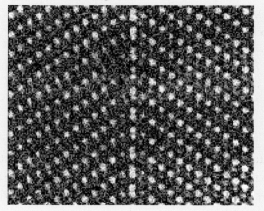

2 Sensing

Sensors, often miniaturised, are used to make measurements, monitor production processes, and control machines. We will give examples of:

● making microsensors

● measuring and controlling signals from sensors

● using sensors for light, temperature, movement and strain

The examples will lead you to new ideas about electric current and potential difference, and about building electric circuits.

2.1 Making very small things

On a November day in 1960, engineer William McLellan carried a large grocery carton into a laboratory at the California Institute of Technology. The year before, in a talk 'There's plenty of room at the bottom', the physicist Richard Feynman had offered a $1000 prize to anyone who could make an electric motor no bigger than half a millimetre on a side. The box looked absurdly big to hold such a motor. But what McLellan took from the box was not a motor but a microscope; through the microscope Feynman watched McLellan's tiny motor turning. Feynman wrote a cheque.

Two years earlier, in 1958, another engineer, Jack Kilby of Texas Instruments, had written in his laboratory notebook,

> *Extreme miniaturisation of many electrical circuits could be achieved by making resistors, capacitors and transistors & diodes on a single slice of silicon.*

Kilby and others filed patents for making these miniaturised circuits—the first integrated circuits. Feynman's vision, of the huge potential of making tiny circuits and tiny machines, was already coming true. The era of microtechnology had begun.

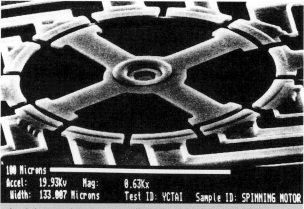

A miniature electric motor. This motor, smaller than the one William McLellan showed Richard Feynman, is the thickness of a human hair across (0.1 mm).

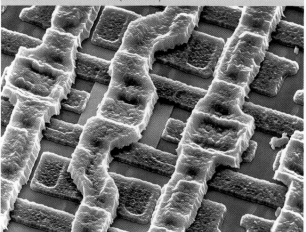

False colour image of a computer memory chip. Pathways for electric currents criss-cross the chip. The conducting strips are 3 μm wide. Today they would be ten times narrower.

A flexible bar of silicon, carrying a sharp tip. The bar is about 20 μm wide and about 5 μm thick. The tip is at most 1 μm across. The bar is part of an atomic force microscope, which scans the tip over a surface. The bar bends as the tip is attracted to the surface by intermolecular forces. The bending indicates the height of the surface below the tip. Objects down to the scale of molecules can be resolved.

Today, millions of microchips run unseen inside personal computers, calculators, domestic appliances and motor vehicles. Richard Feynman was right: there is plenty of room at the bottom.

Miniature circuits and devices have completely changed the world of measurement and control. A family car today will contain hundreds of microsensors, each monitoring how the car is performing. One of them, no larger than a printed letter on this page, sits in the steering column checking continually on whether the car is decelerating rapidly in a crash. If that happens, the microsensor triggers the release of air bags to protect driver and passengers.

Go into a hospital intensive care ward. Microsensors attached to the patient monitor breathing, blood pressure, heartbeat and other vital functions. Signs of trouble quickly summon nurses and doctors to help. By being so small the sensors can be unobtrusive and convenient. Tiny microsensors can help watch over delicate premature babies without causing them unnecessary stress. A microsensor, with communications attached, may also be implanted under a patient's skin to warn of danger signs. Others, attached to animals in the wild, signal movements of herds of deer or flocks of birds.

One way to make these tiny objects out of silicon and other materials is alternately to deposit material on a silicon surface and then to selectively cut away parts with acids and other chemicals. That is how the miniature silicon bar used in an atomic force microscope, shown on page 29, was made. Most microchips are also made in this way.

A very different way to shape miniature objects is to bombard them with a beam of fast-moving charged particles. It works very much like using sand-blasting to strip paint and rust from car bodies. Rapidly moving ions 'blast away' surface atoms, knocking them out of the surface, making it possible to smooth or shape a surface right down to the near-to-atomic scale. Diamond tips used as super-sharp blades for surgery can be made like this. The process is called 'ion-beam machining'.

● Images of micromachined structures

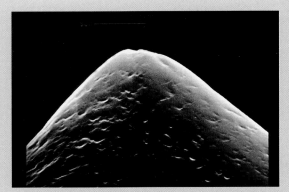
A diamond stylus finished to a sharp point. Here you see the result of mechanical polishing. The tip is sharp to about 5 μm.

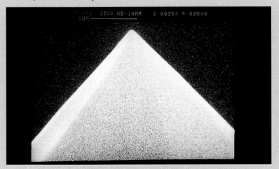
Here you see the much sharper tip, about 0.1 μm across, produced by bombarding the surface with energetic ions.

Machining a tip with a beam of ions

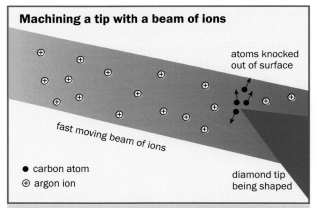

atoms knocked out of surface

fast moving beam of ions

● carbon atom
⊕ argon ion

diamond tip being shaped

Machining with a beam of ions. Argon ions are made by firing electrons at argon atoms. The ions are then accelerated into a beam, which strikes the workpiece and knocks atoms out of the material. The shape is controlled by varying the angle of the surface to the beam of ions. Material can be removed at a rate of up to 0.1 μm per minute. The process is like using sand-blasting to clean surfaces, but on an atomic scale.

Beams of moving ions

A beam of moving charged particles carries an electric current without wires. The beam must travel in a vacuum, otherwise it will be scattered by colliding with gas molecules. Other examples of moving charged particles are:

- the beam of electrons which lights up the screen of your television set
- a beam of protons in a particle accelerator used to probe the nature of matter
- charged particles coming from the Sun causing the 'Northern lights' (Aurora Borealis)

Such a beam carries energy because the particles are moving. The moving particles do what moving particles usually do: they knock into things, delivering energy to a target. The beam carries an electric current because the particles are electrically charged and moving.

Calculations on ion beams and currents

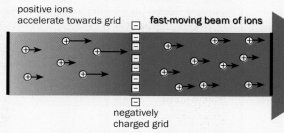

positive ions
accelerate towards grid

fast-moving beam of ions

negatively
charged grid

charge on ion = q electric current $I = Nq$ number N ions arrive per second

How many ions hit an atom every second in this beam?

Typical beam data:
beam diameter 10 mm
(beam area = 100 mm^2 approx. = 10^{-4} m^2)
beam current I = 20 mA = 20×10^{-3} A

Data about the ions:
charge on argon ion = $+1.6 \times 10^{-19}$ C

1 Number of ions in beam:

Number of ions arriving per second $N = \dfrac{I}{q} = \dfrac{20 \times 10^{-3}\ \text{A}}{1.6 \times 10^{-19}\ \text{C}}$

$$= 12 \times 10^{16}\ \text{s}^{-1}$$

$$= 10^{17}\ \text{s}^{-1}\ \text{approx.}$$

2 Area covered by one atom:

Area covered by one atom = $(0.1\ \text{nm})^2$

$$= 10^{-20}\ \text{m}^2$$

3 Number of atoms in area of beam:

$$= \frac{\text{area of beam}}{\text{area of atom}} = \frac{10^{-4}\ \text{m}^2}{10^{-20}\ \text{m}^2} = 10^{16}$$

4 Number of ions hitting an atom per second:

$$= \frac{\text{number of ions per second in beam}}{\text{number of atoms in beam}}$$

$$= \frac{10^{17}\ \text{s}^{-1}}{10^{16}} = 10\ \text{ions per second}$$

The very large and very small numbers come together to give a very simple and remarkable result:

In this beam an atom is bombarded by about 10 ions each second.

- Help with calculating current and power in an ion beam

- Calculating with powers of ten and logarithms

Potential difference

What makes charged particles move? It is other charged particles attracting or repelling them. The positive argon ions used for ion beam machining are accelerated by being pulled towards a concentration of negative charge on the grid.

In the early 1800s, alone in the Nottinghamshire countryside the young self-taught mathematician George Green was thinking about the energy of electric charges. He thought of a very simple way to work out which way a charge will be pushed or pulled by concentrations of other charges. He realised that you do not need to know where all these other charges are. All you need to know is the potential energy of the charge. The charge will be pushed or pulled in the direction in which the potential energy decreases.

This is just like a massive object on a hill. Going uphill, the potential energy increases as the mass gets further from the Earth—a concentration of mass. Going downhill, the potential energy decreases. And the mass is of course pulled downhill by gravity. George Green's idea was that charges go downhill too: down 'electrical hills'.

Green gave the name 'electric potential' to the potential energy divided by the charge. Between two places there can be a difference in the electric potential: a **potential difference**. Charges go down drops of potential.

Having thought of all this by himself, George Green became a student at the late age of 40. Not surprisingly his Cambridge College soon made its student a Fellow. His ideas (Green's functions) are still very important in modern theoretical physics.

Potential differences are measured in **volts**. One volt is one joule of energy per coulomb of charge. A thundercloud may generate a potential difference of millions of volts.

Examples of potential difference

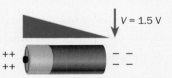

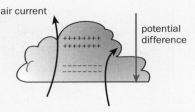

A dry cell. The chemical reaction drives electrons to one pole leaving positive charge at the other. The difference in distribution of charges creates a potential difference between the poles.

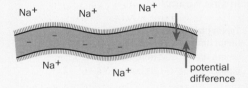

A storm cloud. Strong air currents rub ice crystals against one another, separating charges. The potential difference between top and bottom of the cloud can be millions of volts.

A nerve cell. 'Pumps' in the walls of the cell drive sodium ions outside the cell. This difference leads to a potential difference of about 70 mV between inside and outside. Changes in this potential difference provide the electrical signals for our nerves.

Uneven distributions of charge make potential differences

A flash of lightning, together with its spectrum, obtained by putting a diffraction grating in front of the camera lens. The red colour is from hydrogen, the green from nitrogen, and the blue from oxygen.

Energy and power in a beam of ions

Beams of moving ions have many uses. Ions accelerated by going round and round inside a cyclotron are used to make radioactive tracers for medical investigations. Ion beams are used to 'implant' ions in silicon to make it conduct electricity better. An ion beam engine firing a jet of xenon ions has been proposed for accelerating space craft making interplanetary journeys. Focused beams of ions are used to cut out shapes in tiny devices, in constructing micromachines.

Such tasks require energy. In a beam accelerated by a potential difference, an ion gains kinetic energy. Ion beams carry energy.

An ion beam from an accelerator. The beam has come out into the air, and makes a blue glow because it ionises the air molecules, which give out light as they recombine.

Calculating current, energy and power

Basics

Electric current is the rate of flow of charge

$$I = \frac{\Delta Q}{\Delta t}$$

Electric potential difference V between two places is the difference in potential energy ΔE of a charge ΔQ going between those places, per unit of charge

$$V = \frac{\Delta E}{\Delta Q}$$

Power P is the rate of delivery of energy E

$$P = \frac{\Delta E}{\Delta t}$$

Symbols and units

	units
Q = charge	coulomb C
I = current	ampere A
V = potential difference	volt V
E = energy	joule J
P = power	watt W
t = time	second s

The symbol Δ

The symbol Δ means 'change in'. It always goes with another quantity, for example ΔQ is 'change in charge'. It never stands for anything by itself.

Calculating formulae

Energy E given to charge Q on an ion going through a potential difference V

$$E = QV$$

If N ions arrive per second, then:
Current I in beam is charge delivered per second

$$I = NQ$$

Power P delivered by beam is energy delivered per second

$$P = NE = NQV$$

Power delivered by a current I going through a potential difference V

$$P = IV$$

Try these

The magnitude of the charge on an electron, or on a singly charged ion, is 1.6×10^{-19} C

1   To collect a charge of 10 μC, for how long should an electron beam carrying a current of 1 μA be switched on?

2   A particle accelerator delivers a million singly charged ions to a target every microsecond. What is the beam current?

3   The electron beam in a television set is given energy by travelling across a potential difference of 10 kV. What is the beam current if the rate at which energy reaches the screen is 50 J s^{-1} (i.e. 50 W)?

4   The energy given to an electron as it goes from one terminal to the other of a battery is 2.4×10^{-19} J. What is the potential difference generated by the battery?

5   An electron given energy by travelling across a potential difference of about 20 V can knock another electron out of an argon atom. What energy does it have before the collision?

6   An electric kettle is labelled '230 V, 2.3 kW'. What current does it draw?

Answers 1. 10 s **2.** 0.16 μA **3.** 5 mA **4.** 1.5 V **5.** 3.2×10^{-18} J **6.** 10 A

YOU HAVE LEARNED

- That sensors and circuits can be made on a scale of micrometres.

- That an electric current consists of moving charged particles.

- Electric current $= \dfrac{\text{charge transferred}}{\text{time taken}}$; $I = \dfrac{\Delta Q}{\Delta t}$

- Potential difference $= \dfrac{\text{potential energy difference}}{\text{charge transferred}}$; $V = \dfrac{\Delta E}{\Delta Q}$

- Power $=$ current $\times$ potential difference; $P = IV$

electric charge, electric current, potential difference, ion, electrical power

2.2 Miniature circuits

"...there are now more transistors made every year than raindrops falling on California, and it costs less to produce one than to print a single character on this page."

Michael Riordan and Lillian Hoddeson, in *Crystal Fire* (1997) W. W. Norton

The factories of the world-wide microchip industry which build tiny electric circuits in silicon are places obsessed with purity and cleanliness. Silicon crystals with much less than one part in a billion of impurities are grown. Then, layer after layer, the silicon is 'doped' with other elements, new layers of material are evaporated onto the surface, protective layers are added and then cut away selectively, and chemicals are used to etch shapes into the material.

The tiniest particle of dust can ruin the emerging circuits. One flake of dandruff spells disaster. So the chips have to be made in ultraclean environments. The air in the room must be filtered, its doors need air-locks, and people must wear special clothing.

A picture of the unseen dirt around us (false colour image). The dust mite in the middle, obliged to scramble amongst dust boulders, is about 0.1 mm long.

Charges moving in conductors

A wire carrying a current does not look as if anything is going on inside. But charges are indeed on the move in it. The drift of charges when a current flows can be seen in liquids, if the current is carried by ions that colour the liquid.

In metals, electric currents are carried by moving electrons. One good reason to think that there are electrons inside metals is that electrons 'boil' out of metals when you heat them. You do this every time you turn on a television set. A heated oxide-coated cathode gives out electrons, which are formed into a beam and accelerated towards the screen.

In electroplating with a solution of copper sulphate, current is carried by positive copper ions and by negative sulphate ions. The ions move from one electrode to the other. In red neon advertising signs, current is carried by neon ions.

The mobile charges which carry currents in silicon are mostly put there by the chip-maker. They are put there by 'doping' the silicon with phosphorus or boron. See chapter 5 for more about how materials conduct electricity.

These beautiful photographs, taken at one second intervals, show a blue colour spreading in a heated transparent crystal of potassium chloride. There is a potential difference of about 300 V between the needle and the support of the crystal. Electrons are injected into the crystal from the needle. Trapped in spaces where an ion is missing in the crystal, the electrons absorb yellow light and the crystal looks blue. You can see the colour spreading as electrons move through the crystal.

Getting charge carriers moving

Like molecules in a gas, the mobile charge carriers in a conductor move about continually in all directions. For a current to flow, they have to be made to move together in one direction (on average). It is a potential difference which gets this drift in one direction going.

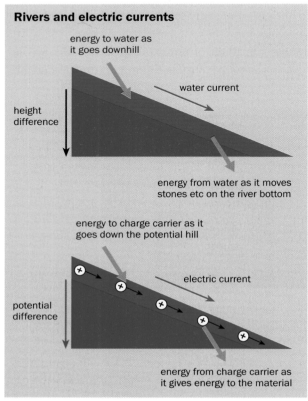

Rivers and electric currents

energy to water as it goes downhill

water current

height difference

energy from water as it moves stones etc on the river bottom

energy to charge carrier as it goes down the potential hill

electric current

potential difference

energy from charge carrier as it gives energy to the material

An electric current is something like a river. The water in a river does not necessarily speed up as it goes downhill. Energy gained by the water from going downhill is given up again as the moving water forms eddies and rubs against the river bed. Energy leaves the moving water as fast as it enters.

The same happens with electric currents. In a conductor the drift of mobile charge carriers is steady. They gain energy by 'falling down' a potential difference, and give it up again to the conducting material in which they are moving. Energy flows in from the potential difference, and out again to heat up the conductor.

The charge carriers in a conductor do not all start moving the instant the potential difference is switched on. It is easy to show that a signal takes about 1 μs to travel the length of a 300 m reel of cable. When the potential difference is switched on, an electromagnetic impulse sweeps round the circuit, leaving behind potential differences along the wires which keep the charge carriers moving. Computer designers have to worry about these delays: hundreds of calculations may be over and done with in 1 μs. Reducing delays is one reason to build microchips small.

How much current from a potential difference?

The designer of a microchip needs a quantity which says how much current flows for a given potential difference. The amount differs from one conductor to another. Some, with many charge carriers which move easily, give a large current for a small potential difference. Others, with few charge carriers able to move, only carry a small current. The conductance G measures how well a wire, a chip component or an electroplating cell conducts:

$$\text{conductance } G = \frac{\text{current } I}{\text{potential difference } V}$$

The unit of conductance is amperes per volt, or siemens, symbol S.

The conductance of an object is not necessarily the same whatever the current or potential difference. The conductance would increase if the moving charges started to hit and ionise other atoms in the material, so providing more mobile carriers.

Circuit designers also think in the opposite way, of how badly things conduct. This is the resistance.

$$\text{resistance } R = \frac{\text{potential difference } V}{\text{current } I}$$

$$\text{resistance } R = \frac{1}{\text{conductance } G}$$

The unit of resistance is volts per ampere, or ohm, symbol Ω.

If you know the conductance, you can immediately work out the resistance. Told the resistance, you can immediately work out the conductance. It's like switching from thinking about the speed of the car to thinking about how long the trip will take. Pairs of reciprocal quantities like these are common in physics.

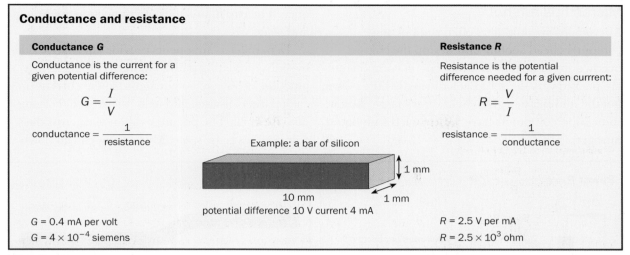

Conductance and resistance

Conductance G	Resistance R
Conductance is the current for a given potential difference:	Resistance is the potential difference needed for a given currrent:

$$G = \frac{I}{V}$$

$$\text{conductance} = \frac{1}{\text{resistance}}$$

Example: a bar of silicon

1 mm
10 mm
1 mm
potential difference 10 V current 4 mA

$$R = \frac{V}{I}$$

$$\text{resistance} = \frac{1}{\text{conductance}}$$

G = 0.4 mA per volt

G = 4×10^{-4} siemens

R = 2.5 V per mA

R = 2.5×10^3 ohm

Words matter

There are a lot of new words to learn in studying electricity. These words have a long history, and were sometimes chosen when people did not understand electricity very well. 'Potential difference' strikes us as a term George Green chose well; a difference between two places. But a less good term still in use for nearly the same thing is 'electromotive force', which is not a force, but is energy divided by charge.

'Current' seems to us another good term: it says, 'something is flowing'. But the symbol *I* for current goes back to the time when current was called 'electrical intensity'; 'intensity' sounds nothing like a flow.

Michael Faraday chose the word 'ion', taking it from a Greek word meaning 'traveller', when he wanted a word to make people believe that the particles in conducting solutions actually do travel from one electrode to the other.

'Conductance' helpfully suggests the ability to conduct, but the more common term 'resistance' may wrongly suggest that wires 'fight back' against a potential difference, when they may merely lack charge carriers which can move.

The origin of the choice of the word 'charge' itself we do not know. Was it by analogy with a charge of gunpowder? Who knows? Whatever their origins, finding the right words is a serious matter in physics. And getting things clear may take time.

Ohm's law

The conductance or resistance can be calculated at any given current or potential difference. If the current or potential difference changes, the conductance and resistance may change.

If the conductance and resistance are constant, independent of the current or potential difference, the conductor is said to obey Ohm's law, or to be 'ohmic'.

Ohm's 'law' thus says that the conductance, and resistance, of a given component is constant. The same value can be used in calculations whatever the current or potential difference.

Most metals are ohmic at constant temperature; ionised gases are not.

- Conductance and resistance in a filament lamp
- Conductance and resistance in a neon lamp
- Conductance and resistance in a silicon diode
- Conductance and resistance in an ohmic resistor

Microchips run hot

Microchips in computers often run hot, and computers need cooling fans to stop the chips overheating. The reason is that circuit components are packed very densely. Each dissipates little power, but the number in a given area is very large. Result: some supercomputers have to be cooled with liquid nitrogen.

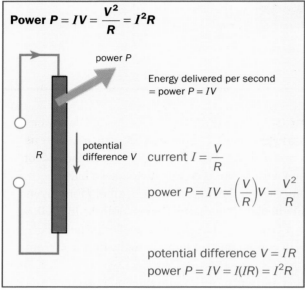

$$\text{Power } P = IV = \frac{V^2}{R} = I^2R$$

power P

Energy delivered per second
= power $P = IV$

R

potential difference V

current $I = \dfrac{V}{R}$

power $P = IV = \left(\dfrac{V}{R}\right)V = \dfrac{V^2}{R}$

potential difference $V = IR$
power $P = IV = I(IR) = I^2R$

For other purposes—heating and lighting—the power dissipated when a potential difference makes a current flow is just what is wanted.

Connecting conductors together

Connections on a silicon chip can be very complicated. The same source of potential difference may have to send current through several components at once. The two main kinds of connection are *parallel* and *series*.

In parallel, components provide alternative side-by-side conducting paths. Charges coming in to a junction all have to go out again, so the currents in each parallel branch add up to the current coming in to them all. The potential difference across the components is the same. More water can go down

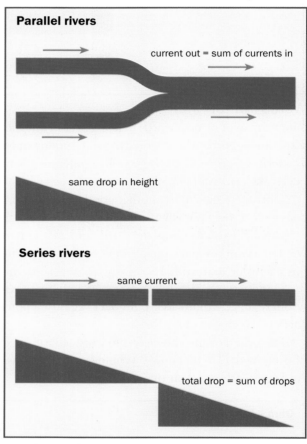

Parallel rivers

current out = sum of currents in

same drop in height

Series rivers

same current

total drop = sum of drops

two rivers side by side on the same hill than can go down one. Thus for components in parallel, the *conductances add up*.

In series, components are connected one after the other, and each requires a part of the potential difference to drive charge through it. The current is the same in both but has 'further to go'. A greater drop is needed for a longer river to carry the same flow of water. Thus for components in series the *resistances add up*.

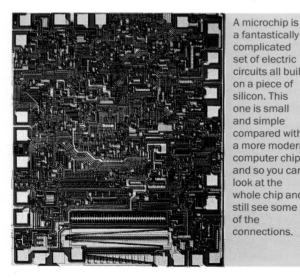

A microchip is a fantastically complicated set of electric circuits all built on a piece of silicon. This one is small and simple compared with a more modern computer chip, and so you can look at the whole chip and still see some of the connections.

● Scaling-down arguments about why microchips run hot

Conductors in parallel and series

Parallel

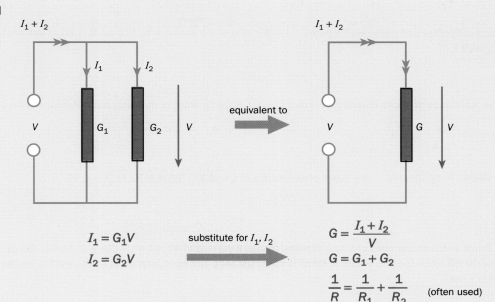

$$I_1 = G_1 V$$
$$I_2 = G_2 V$$

substitute for I_1, I_2

$$G = \frac{I_1 + I_2}{V}$$
$$G = G_1 + G_2$$
$$\frac{1}{R} = \frac{1}{R_1} + \frac{1}{R_2} \quad \text{(often used)}$$

Currents add up, potential difference is the same for both ⟶ **conductances add up**
Example: lamps in domestic wiring circuit

Series

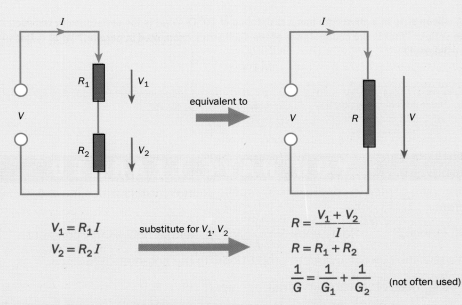

$$V_1 = R_1 I$$
$$V_2 = R_2 I$$

substitute for V_1, V_2

$$R = \frac{V_1 + V_2}{I}$$
$$R = R_1 + R_2$$
$$\frac{1}{G} = \frac{1}{G_1} + \frac{1}{G_2} \quad \text{(not often used)}$$

Potential differences add up, current is the same for both ⟶ **resistances add up**
Example: potential divider (see section 2.3)

● Kirchhoff's laws: Current and potential
difference in complex circuits

Try these

1 The conductance of a wire with a potential difference of 5 V across it is 250 mA V^{-1}. What current flows through it?

2 The resistance of a wire carrying a current of 0.1 A is 25 Ω. What is the potential difference across it?

3 A silicon component in a chip has a conductance of 10 μA V^{-1}. What is its resistance?

4 A flame between the ends of a pair of wires carries a current of 10 mA when the p.d. across the wires is 100 V, and 40 mA when the p.d. is 200 V. What is the resistance and conductance in each case? Does the flame obey Ohm's law?

5 A typical 60 W mains light bulb carries a current of 0.25 A. How much current is drawn if five such lamps are connected together in parallel?

6 A silicon strip in a microchip has a resistance of 10^4 Ω. What is the resistance of ten such strips connected in series? What is the conductance of ten such strips connected in parallel? What is this expressed as a resistance?

Answers 1. 1.25 A **2.** 2.5 V **3.** 10^5 Ω **4.** $R = 10^4$ Ω, $G = 10^{-4}$ S or 100 μA V^{-1}; $R = 5 \times 10^3$ Ω, $G = 2 \times 10^{-4}$ S or 200 μA V^{-1}; the flame does *not* obey Ohm's law **5.** 1.25 A **6.** 10^5 Ω; 10^{-3} S; 10^3 Ω

YOU HAVE LEARNED

- Currents are carried in conductors by mobile charge carriers which move under a potential difference.
- Conductance $G = \dfrac{I}{V}$; Resistance $R = \dfrac{V}{I}$; $R = \dfrac{1}{G}$ (neither necessarily constant).
- Ohm's law: the current is directly proportional to the potential difference.
- The resistances of resistors in series add up: $R = R_1 + R_2$
- The conductances of resistors in parallel add up: $G = G_1 + G_2$, and hence $\dfrac{1}{R} = \dfrac{1}{R_1} + \dfrac{1}{R_2}$

ion, conductance, resistance, parallel circuit, series circuit, Ohm's law, electrical power

2.3 Controlling and measuring potential differences

Radio and hi-fi

To turn up the volume on the radio, you turn a knob, its volume control. On an old radio there may be a scratchy sound as you turn the knob. What is going on behind? The signal from the radio is controlled by 'tapping off' part of it with a sliding contact moving along the surface of a high resistance. The high resistance may be a film of carbon or a wire coil. The device is called a **potential divider**, sometimes 'potentiometer', or even 'pot' for short. It works by having the signal from the radio—a varying potential difference—across the whole resistance, but the signal to the radio loudspeaker taken from across only a part of the whole resistance. The scratchy noise on an old radio may come from dirt on the surface of the resistance, which briefly spoils the contact of the slider as it moves. Really top quality high-fidelity equipment sometimes uses a chain of fixed resistors in between gold-plated contacts to beat the noise problem.

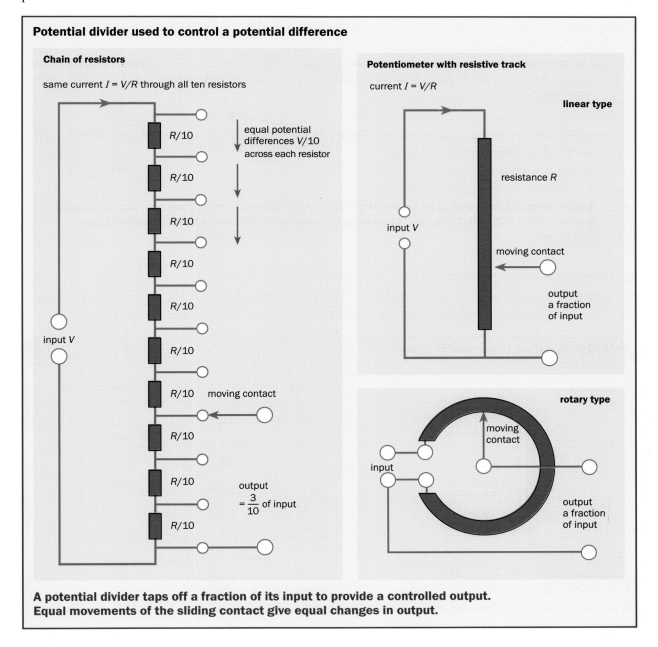

Potential divider used to control a potential difference

Chain of resistors

same current $I = V/R$ through all ten resistors

$R/10$ — equal potential differences $V/10$ across each resistor

$R/10$

$R/10$

$R/10$

$R/10$

$R/10$

$R/10$ moving contact

$R/10$

$R/10$ output $= \dfrac{3}{10}$ of input

$R/10$

input V

Potentiometer with resistive track

current $I = V/R$

linear type

resistance R

input V

moving contact

output a fraction of input

rotary type

moving contact

input

output a fraction of input

A potential divider taps off a fraction of its input to provide a controlled output. Equal movements of the sliding contact give equal changes in output.

In the top quality volume control all the resistors are in series. They carry the same current. If their resistances were all the same, the potential differences across each would be all the same. Potential differences add up along a series circuit, so the potential difference across say three resistors, if there are ten in all, is three tenths of the potential difference across them all. If the potential divider is a continuous carbon film, each piece of film is still in series with the next, and the same idea of tapping off a fraction of the total potential difference still works. Now the output potential difference is proportional to how far the slider has been moved along—or around—the potentiometer.

Car fuel gauge

The fuel gauge of a car does not actually measure the amount of fuel in the tank. Instead, it measures something easier, the *level* of fuel in the tank.

A fuel gauge can use a potential divider to detect the position of the float. Petrol is used up—the float goes down—the tilting float arm turns—the sliding contact inside a potential divider moves. Result: the output of the potential divider changes and a signal goes to the fuel gauge. All too soon the driver is looking for a petrol station. In a similar way, a position sensor can keep track of how far a robot's finger is from what it is touching, and an angle sensor can measure the angle of bending of the robot's elbow.

In a sensor measuring the angle of a robot arm you want equal output for equal changes in angle. A potentiometer with a uniform resistive track can give an output proportional to the distance moved by the sliding contact. A graph of output potential difference against slider position is a straight line. The relation is *linear*. For linear changes, translating back from a change in output to the change in position is very simple—which is why linear devices are useful and important.

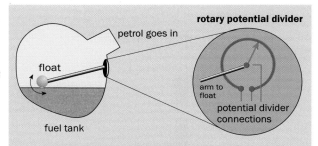

Fuel gauge in a car. A float connected to an arm rises or falls with the fuel level, and the arm turns, moving a contact round a resistor in the form of a circle. The resistor is a potential divider acting as a sensor.

Non-linear changes: calibration

You want to read straight from the fuel gauge how much petrol there is in the tank. If you look, you'll see that many fuel gauges do not move round the dial by equal amounts for equal numbers of litres of petrol

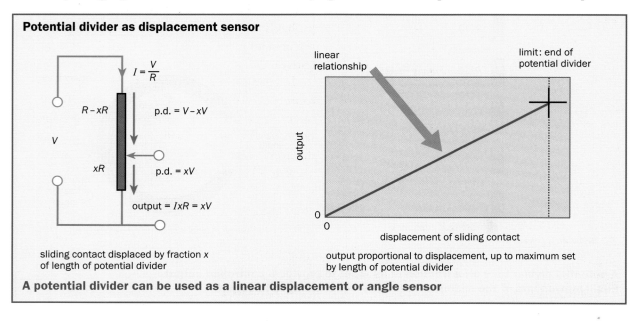

A potential divider can be used as a linear displacement or angle sensor

used up. That is, the scale is often very *non-linear*, and drivers sometimes think they have more left than is really there. The translation back from output to input is not now so simple.

One reason is that petrol tanks in cars often have to be built in complicated shapes to fit into the space available. The fuel level certainly goes down as petrol is used up, but changes in the level have no simple relationship to the amount of fuel used. Thus the relationship between float level and amount of fuel is very non-linear. Its graph would have a strange shape—certainly not a straight line. Such a graph is called a *calibration curve*. You will need one for any sensor you make or buy. A calibration curve tells you how to look up the input to a measurement system if you know its output. Calibration curves are nowadays often built right into sensor systems, as 'look-up tables' relating input to output.

'Look-up tables' are especially useful when a sensor is non-linear. This is part of a general trend: software replacing hardware. Mass producing large numbers of carefully designed hardware items is expensive; copying software, once it is written, is extremely cheap.

Results can fluctuate

Noise, like that from a dirty volume control, is often present in measurements. For example, the level of

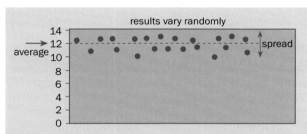

Results can vary about the 'true' value. If the variations are random, their average will be a better indication of the 'true' value than any one result alone. But errors need not be random. They can be systematic, consistently wrong in one direction, or they can be accidental.

The average of a set of results varies less than individual results vary.

fuel in a petrol tank will vary as the fuel sloshes around. A sluggish gauge might not notice; one with a fast response time would need to take frequent sample readings and calculate their average over time.

This is an example of a very general problem in instrumentation: results can vary around the 'true' value in an unsystematic way. Some other examples:

- radioactive carbon dating in archaeology: the number of decays recorded in a given time varies randomly
- noise in astronomical images (see chapter 1)
- fluctuations in world-wide temperatures making it hard to see if the 'greenhouse effect' is actually raising the average temperature.

Noise in images can be reduced by averaging values of neighbouring pixels (chapter 1)—fundamentally the same idea as averaging readings which fluctuate. Averaging works well if fluctuations are unsystematic. The average varies less than any one result.

Small variations are usually most important for sensitive instruments. When the physicist R V Jones, who led much of the scientific effort during the Second World War, built an extremely sensitive tilt-detector in the University at Aberdeen, he found that he could detect the building tilting as the tide came in.

These examples show that errors or variations in readings are sometimes random in nature, as with radioactive decays, and sometimes systematic, as when the tide comes in and the ground tilts under its weight. The sloshing of petrol in the tank is probably not really random (it is quite like the tide going in and out) but it is complicated and unsystematic.

In your own experiments there are many such sources of variation. Watching a digital meter you often see the last digit varying. Repeated readings rarely agree exactly. It will often be useful to report the average (or sometimes better, the median) of a set of results, together with an estimate of their spread.

How to... Handle data:
- Dealing with variability in data

- Beta Pictoris: A 'noisy' astronomical image of a possible dust cloud round a star

Try these

1 A potential divider 50 mm long has a p.d. of 2 V across it. The sliding contact is moved 5 mm along the potential divider. By how much does the output of the potential divider change? The slider is now moved a further 10 mm. What is the change in output?

2 A potential divider has a total resistance of 1 kΩ, and carries a current of 3 mA. What is the output p.d. if the sliding contact is at its centre point?

3 The swinging of a pendulum is detected by a circular rotary potential divider. The output seen on an oscilloscope has a variation from peak to trough of 100 mV. The input to the potential divider is 1 V. Estimate the angle through which the pendulum is swinging.

4 In use a potential divider becomes hot and its resistance increases from 10 kΩ to 11 kΩ. If the input p.d. is 2 V, by how much does the current through it change? By how much does its output p.d. change if the sliding contact does not move?

5 Sketch a V-shaped petrol tank. Use your sketch to construct a look-up table to convert levels in the tank to volumes in the tank.

6 If you measure the resistances of each of a large batch of supposedly identical commercial resistors, you get a spread of results. If you measure the intensity of pixels in an area of an image which is supposedly uniformly bright, you get a spread of results (chapter 1). Explain the connection between averaging a set of experimental results and smoothing an image.

Answers 1. 0.2 V, 0.4 V **2.** 1.5 V **3.** 36 degrees; actually less because the potential divider cannot fill a complete circle
4. 0.02 mA, the output does not change

YOU HAVE LEARNED

- How a sliding contact potential divider works as a sensor for displacement.

- That a sliding contact potential divider can be a linear device (output p.d. proportional to displacement).

- That look-up tables can be used to deal with some non-linear relationships.

- That unsystematic, perhaps random, fluctuations can be reduced by averaging.

A-Z | potential divider, linear device, linear, random error, average, mean, median

2.4 Sensors and our senses

Would you like to live in a 'smart' home which recognises you when you come in, and puts on the kettle for you? Perhaps intelligent surveillance systems are more likely to get here first. What sensing systems would they need? Sight? Hearing? Touch? Smell? Taste? All are possible electronically.

Cars are now largely built by robots. A few people are still needed but only to look after the robots. The robots must be able to sense, to fractions of a millimetre, how accurately they are putting parts of the car in place. The cars they build contain hundreds of sensors, for example to detect and prevent skidding. In these uses what do sensors need to measure or detect? Position? Speed? Acceleration?

Information about the world can be obtained in many ways. Your senses use only a few of them, but have the advantage of being made smart by being connected to a powerful brain. Modern sensors pick up information that you cannot sense, but have only recently started to get even a little 'smart'. For example, a sensor may measure the size of an interfering effect such as temperature, and compensate for it.

● An electronic toaster which senses when the toast is done

Electronic ears

A microphone is an electronic ear, producing a varying electrical signal from pressure changes caused by sounds. In chapter 1 (page 3) you can see the use of a piezoelectric crystal such as quartz to detect high frequency (ultrasonic) sound. With a piezoelectric gas lighter you squeeze and get a spark. Both generate their own potential difference. Such crystals usually have a very high resistance indeed. Almost no current flows in a circuit connected to a high resistance source like this.

Electronic skin

Your skin detects changes of temperature, as well as of pressure. One way to detect temperature changes electrically is to use a thermocouple. A thermocouple is just a pair of different metals joined together:

Radiation sensor. The sensor is a tiny 3 mm by 3 mm chip. Infrared radiation falls on the chip and is absorbed by a thin 0.55 mm diameter black disc at the centre of the chip, warming the disc. Grids of thermocouples measure the temperature of the disc.

when the junction is made hot a potential difference is created (Seebeck effect). The potential difference is small, of the order of a few millivolts or less. But the resistance of a thermocouple is generally small. So despite the small potential difference, it will give a measurable current.

Different pairs of metals produce different thermoelectric emfs. One of the largest is the 120 μV per degree Celsius from the pair antimony and bismuth. Using multiple junctions with alternating hot and cool junctions increases the effect. Ten antimony–bismuth junctions in series give ten times the emf. You can see four grids of alternating bars of antimony and bismuth in the photograph of the radiation sensor. In this way, the **sensitivity** of the detector can be increased. The sensitivity of a

● Internal resistance: Effect on p.d. from a cell under load

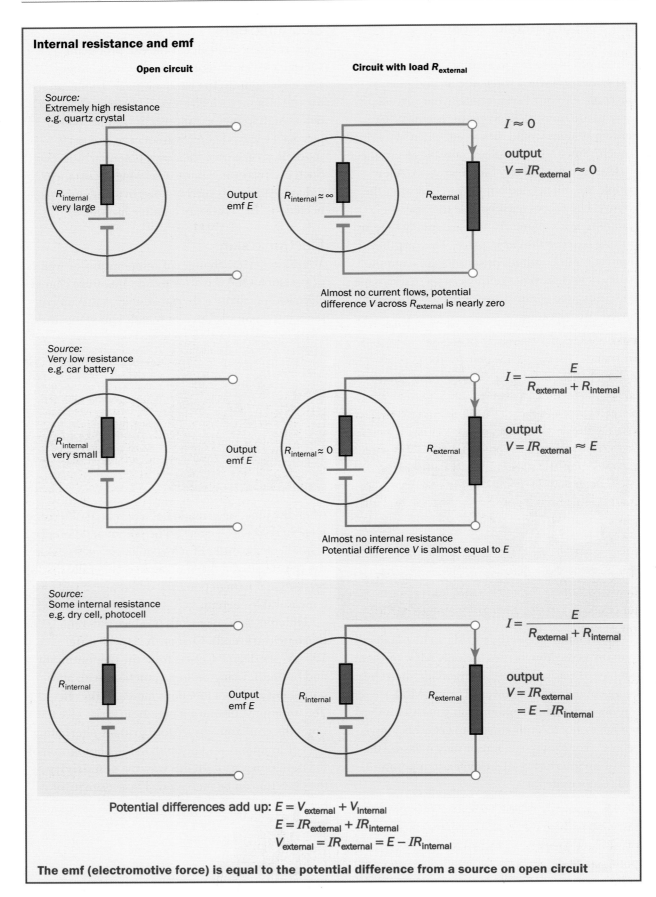

The emf (electromotive force) is equal to the potential difference from a source on open circuit

sensor is the ratio of change in output to change in input, in this case a few mV per degree.

Getting the signal out

The examples of a crystal microphone to detect sound and of a thermocouple to detect warmth show that there can be problems in detecting a signal from a sensor. The thermocouple needs a sensitive voltmeter, able to detect a signal of the order of thousandths of a volt. Too insensitive a detector, and there appears to be no temperature difference. The crystal microphone gives a different problem: the signal may seem not to be detectable because the crystal has a very high resistance. So you connect a detector, but see nothing because not enough current can flow to drive the detector.

This last problem is the problem of **internal resistance** of a source. If connected to a low resistance detector or other load, the potential difference across it may be tiny, because most of the potential difference available is used in driving a small current through the large internal resistance. Only if the detector has itself a very high resistance is there an appreciable potential difference seen as the output of the sensor. Luckily cathode ray oscilloscopes and audio amplifiers can easily be made with a high input resistance, so crystal microphones work well with them.

Electronic eyes

The charge-coupled device in chapter 1 (page 7) is one kind of electronic eye. Another is a silicon p-n junction, which can detect radiation over a wide range of wavelengths, from ultraviolet to infrared.

Notice that in the graph of potential difference against amount of light (illuminance), the amount of light is plotted on a 'times' scale (logarithmic scale). The potential difference increases in equal steps as the amount of light falling on it is multiplied by a constant factor. This more or less matches what the human eye does.

Space craft orbiting the Earth, and the Hubble space telescope, get their power from similar devices, but now made on a grand scale, placed on 'wings' which fold out to catch the maximum of sunlight. Devices which produce an electrical potential difference from light are called photo-voltaic cells.

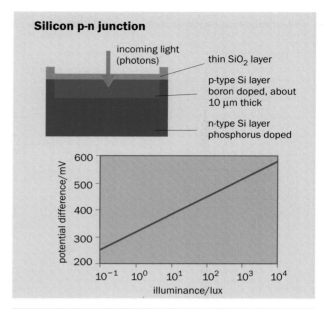

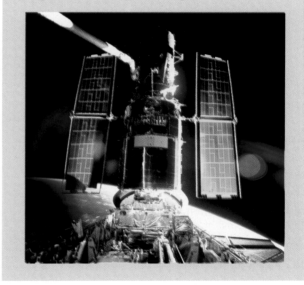

The Hubble telescope, attached to the Space Shuttle, after being repaired and refurbished. The gold coloured wing-like panels are the telescope's power supply: photovoltaic cells which generate electricity from sunlight.

Photoconductive sensors

Materials such as cadmium sulphide (CdS) conduct better when light shines on them. They make sensitive and cheap electronic eyes. They need a power supply, and act by modifying its effect.

To get a large change in resistance, the resistance is increased by laying down the CdS in a long

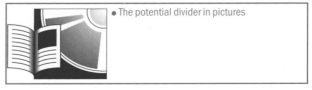

● The potential divider in pictures

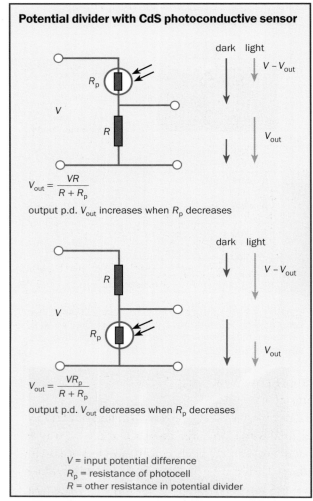

Potential divider with CdS photoconductive sensor

$$V_{out} = \frac{VR}{R + R_p}$$

output p.d. V_{out} increases when R_p decreases

$$V_{out} = \frac{VR_p}{R + R_p}$$

output p.d. V_{out} decreases when R_p decreases

V = input potential difference
R_p = resistance of photocell
R = other resistance in potential divider

zig-zag strip. The change in resistance can be detected using a potential divider. The output of the potential divider can be arranged to increase, or to decrease, when the sensor is illuminated.

Response time

Photoconductive sensors, although sensitive and cheap, are slow. They take about 10 ms to respond to a change in light level; a p-n diode can respond in 1 μs or less. So a photoconductive sensor could not detect rapid changes in brightness, for example from a camera flash gun.

The response time of a thermocouple may be quite large, since the metal junction takes time to warm up. You know the response time of a clinical

response time

thermometer from the time you have to hold it under your tongue. By miniaturising, temperature sensors can be made with response times measured in milliseconds.

Electronic noses

Our noses can detect very small amounts of some substances. Insects detect extremely low levels of chemical signals (pheromones) from potential mates.

A layer of the polymer polypyrrole lying across a narrow gap in a microsensor, deposited there by electrolysis. The conductance of the polymer increases when it is exposed to certain odour molecules, with a sensitivity down to 0.1 parts per million.

Sensors in domestic fire alarms detect smoke. Others detect pollutants, tell fresh food from bad, and are beginning to detect diseases by smelling the breath. Many of these sensors detect a change in resistance caused by a gas in contact with the sensor material. Again, a potential divider can be used to provide an electrical output.

Recently, films of electrically conducting polymers have been used in microsensors. The polymer can be tailored to be sensitive to particular molecules, which is just what is needed for a good 'nose'.

Very thin layers of other kinds of substances, made by floating a film of them on water, are also used in sensors. This way of making layers only a few molecules thick was pioneered in the 1930s by Katherine Blodgett and Irving Langmuir.

Katherine Blodgett was the first woman to obtain a PhD degree in physics at the University of Cambridge, in 1926. As well as pioneering techniques for making what are now called Langmuir-Blodgett films, she played an important role in developing antireflection coatings for lenses, now standard in cameras and binoculars.

● Electronic noses: Telling fresh food from bad and detecting disease in cows

Gas sensors also monitor car exhausts, to check on engine performance. Sensors which change resistance are used here too.

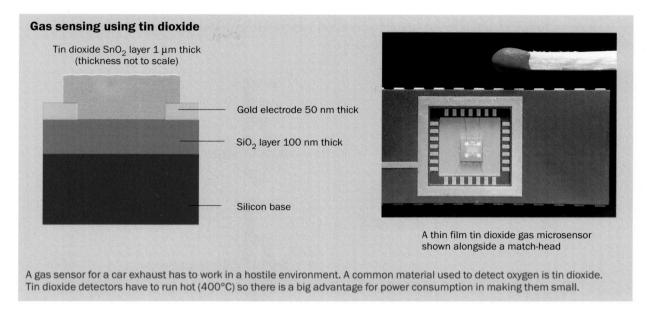

Gas sensing using tin dioxide

Tin dioxide SnO$_2$ layer 1 µm thick
(thickness not to scale)

Gold electrode 50 nm thick

SiO$_2$ layer 100 nm thick

Silicon base

A thin film tin dioxide gas microsensor shown alongside a match-head

A gas sensor for a car exhaust has to work in a hostile environment. A common material used to detect oxygen is tin dioxide. Tin dioxide detectors have to run hot (400°C) so there is a big advantage for power consumption in making them small.

Thermal sensing

The fact that many sensors are affected by changes in temperature means that there are many different ways to sense temperature changes. But it makes problems, too.

Thermistors are beads of metal oxides whose conductance increases very rapidly with temperature. They make very cheap and sensitive temperature sensors, used with a potential divider in the same way as a photoconductive sensor (page 47). Their conductance does not increase linearly with temperature, so it is helpful to use a look-up table (see section 2.3) to convert the p.d. from the potential divider to temperature.

A temperature sensor also helps people with diabetes to monitor their blood sugar levels. The glucose in a tiny drop of blood is oxidised by an enzyme on an organic film; the temperature rise shows the amount of glucose.

Strain gauges

Especially in mining areas, houses are liable to subsidence. The ground below them sags and the house sinks, tilts or twists. But the movement is slow, and it may not be clear whether it is still happening. One way to check is to fit strain gauges to structural parts of the house.

The resistance of a metal increases when it is stretched, simply because it becomes longer and thinner. This gives a way to make a strain gauge.

A strain gauge is a zig-zag strip of metal foil, glued to the surface of the component whose strain is to be monitored. The foil is stretched if the component is stretched. Stretching along the length of the strips in the foil is detected as an increase in resistance. The zig-zag arrangement simply increases the resistance of the gauge—usually about 100 Ω—so as to increase the change in resistance. Such strain gauges are cheap and convenient.

Suppose the bending of a beam has to be measured. Such measurements are needed in investigating the strengths of materials (chapter 4). A really clever trick is to use two strain gauges, so that as the beam bends one is stretched and the other is compressed. Connected in a potential divider, the two gauges give twice the output which one would give alone.

And there is a further bonus: the combination is not so sensitive to changes in temperature. The resistance of a strain gauge increases with temperature. Using one strain gauge in a potential divider, a rise in temperature cannot be told apart from a strain. But using two gauges, if the temperature goes up the resistances of both gauges increase. The output of the potential divider does not change.

This removal of the effect of temperature is an example of something which is very important in measurement. Either the apparatus is designed to eliminate the effect of a disturbing influence, or the effect is measured independently and then allowed for.

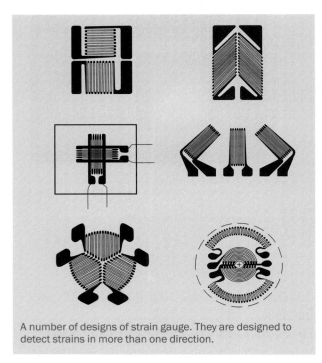

A number of designs of strain gauge. They are designed to detect strains in more than one direction.

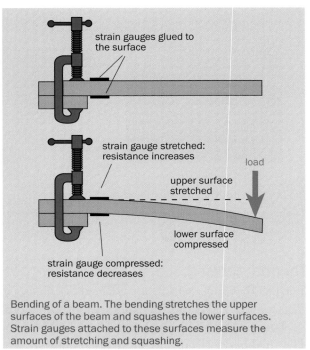

strain gauges glued to the surface

strain gauge stretched: resistance increases

load

upper surface stretched

lower surface compressed

strain gauge compressed: resistance decreases

Bending of a beam. The bending stretches the upper surfaces of the beam and squashes the lower surfaces. Strain gauges attached to these surfaces measure the amount of stretching and squashing.

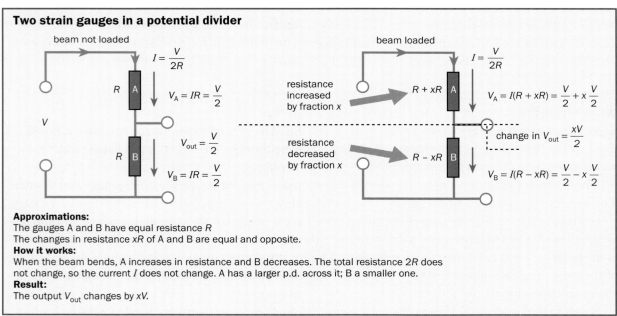

Two strain gauges in a potential divider

beam not loaded

$I = \dfrac{V}{2R}$

$V_A = IR = \dfrac{V}{2}$

$V_{out} = \dfrac{V}{2}$

$V_B = IR = \dfrac{V}{2}$

beam loaded

$I = \dfrac{V}{2R}$

resistance increased by fraction x

$V_A = I(R + xR) = \dfrac{V}{2} + x\,\dfrac{V}{2}$

change in $V_{out} = \dfrac{xV}{2}$

resistance decreased by fraction x

$V_B = I(R - xR) = \dfrac{V}{2} - x\,\dfrac{V}{2}$

Approximations:
The gauges A and B have equal resistance R
The changes in resistance xR of A and B are equal and opposite.
How it works:
When the beam bends, A increases in resistance and B decreases. The total resistance $2R$ does not change, so the current I does not change. A has a larger p.d. across it; B a smaller one.
Result:
The output V_{out} changes by xV.

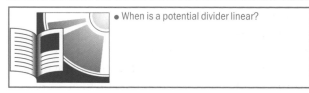

● When is a potential divider linear?

Getting output from sensors

If you are building a sensor system, you have to think how to get a big enough signal out to be recorded by the instruments which you have. There are several problems to think about.

First problem: how big is the signal and how sensitive is the detector? Are you trying to detect a signal of a few millivolts with an instrument that cannot 'see' so small a quantity? Can the signal be amplified?

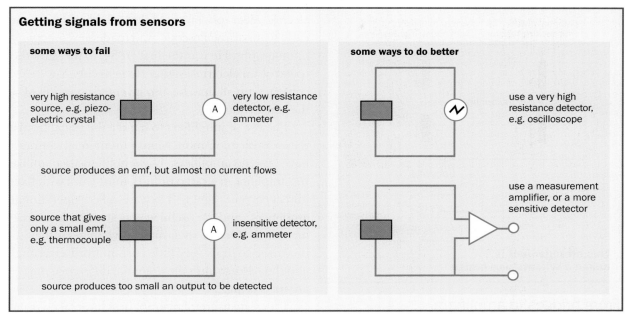

Getting signals from sensors

some ways to fail

very high resistance source, e.g. piezo-electric crystal — very low resistance detector, e.g. ammeter

source produces an emf, but almost no current flows

source that gives only a small emf, e.g. thermocouple — insensitive detector, e.g. ammeter

source produces too small an output to be detected

some ways to do better

use a very high resistance detector, e.g. oscilloscope

use a measurement amplifier, or a more sensitive detector

Second problem: what is the effect of the resistances in the detection circuit? Does the sensor have a large internal resistance? If so, you will get very little current from it. That will not matter only if your detector itself needs very little current—that is, itself has a high resistance.

Internal resistance of source

Many sensors, being made of semiconductors, have quite high internal resistance. Some, like quartz crystals, have very high internal resistance indeed, being practically insulators. When a detector tries to draw current from such a source, the current is limited by the internal resistance of the source. The potential difference across the detector becomes much less than the emf of the source.

Detecting small changes with a potential divider

Suppose you are trying to detect a faint light with a light-sensitive resistor. The feeble light only changes its resistance a little. You have put the light-sensitive resistor in a potential divider. Perhaps the output of about one volt changes by only a few millivolts when the light comes on. How to detect a small change like this?

One solution is to use an electronic amplifier to increase the signal. But it is pointless to amplify the whole one volt signal; you really need to amplify the much smaller change in the output, of only a few millivolts.

How can you get a signal which only contains the *change* in output? A neat way is to compare it with the output of an identical potential divider in which nothing changes. When the light comes on,

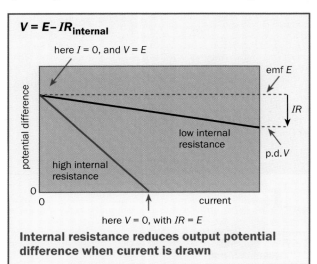

$V = E - IR_{internal}$

here $I = 0$, and $V = E$

emf E

IR

low internal resistance

p.d. V

high internal resistance

here $V = 0$, with $IR = E$

current

Internal resistance reduces output potential difference when current is drawn

one output changes but the other stays the same, and their difference is just the amount of the change —those few millivolts you need to detect. This difference is then well worth amplifying.

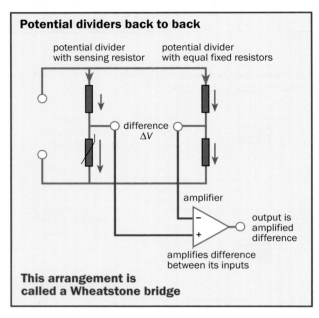

Potential dividers back to back

potential divider with sensing resistor

potential divider with equal fixed resistors

difference ΔV

amplifier

output is amplified difference

amplifies difference between its inputs

This arrangement is called a Wheatstone bridge

From bridges to amplifiers

The idea of putting two potential dividers back to back is the basis of the Wheatstone bridge circuit.

The name of Charles Wheatstone (1802–1875) is remembered by all physicists and electrical engineers for the back-to-back circuit containing four resistors, in which a detector forms a 'bridge' between two potential dividers. But it was not actually his idea at all. It was first devised by Samuel Christie (1784–1865), and then popularised by Wheatstone, who rightly saw its great practical importance. Until quite recently it was fundamental to many measurement circuits, and is still to be found today etched in miniature on many a sensor chip.

The clever idea of the Wheatstone bridge circuit was to make a circuit sensitive only to changes in the ratio of the values of pairs of resistors. So what, you ask? For a long time, resistors were the most accurately made electrical components. So a circuit which compared them was the best for getting

precision results. Equally important, the bridge circuit gives no unwanted output if there is an outside change—like a change in temperature—which affects all the resistances in the same way.

Charles Wheatstone himself did a lot more than just popularise someone else's idea. He built an electric telegraph, invented the 'kaleidophone'—an early form of oscilloscope, and made stereoscopic photographs. He was the first (1834) to measure the speed of an electrical signal along a wire. Not content with all that, he also invented a version of the concertina.

The use of amplifiers to handle the signal from a sensor is very common. Many commercial sensors have on-chip amplifiers. The amplifiers used, unlike the amplifier in a radio or hi-fi, must work with fixed (direct current) signals. They are called 'operational amplifiers', and are useful workaday tools in instrumentation. One of their main advantages is that they need to draw very little current from a circuit, and thus do not alter the potential difference they are measuring.

Today, measurement amplifiers have taken over many of the jobs once done by bridge circuits. Designed to amplify only the difference in potential between two inputs, they too ignore outside changes which affect both inputs equally. You will certainly find yourself using such an amplifier when you build circuits for instrumentation. You might also just find yourself re-inventing Samuel Christie's Wheatstone bridge!

Sir Charles Wheatstone (1802–1875). The Wheatstone Bridge circuit is named after Charles Wheatstone, who popularised its use in measuring devices.

- Find out about the Wheatstone bridge
- Simple circuits for measurement amplifiers

- Effect of load on a potential divider

Try these

1 A micro-thermopile for accurate measurement has been built with 90 junctions in series. Each junction gives 25 µV per degree Celsius. What emf is to be expected for a 10°C temperature difference?

2 A 1.5 V dry cell has an internal resistance of 0.5 Ω. What is the short-circuit current? What is the p.d. across the cell when short-circuited? What is the p.d. across the cell if it delivers a current of 1 A into an external load?

3 A 12 V car battery can deliver a current of at least 48 A into the starter motor. What is the largest value its internal resistance can have?

4 A pair of 10 kΩ resistors are connected as a potential divider, across an input p.d. of 10 mV. What is the p.d. across either resistor?

5 A photoconductive cell has a resistance of 1 MΩ in the dark, falling to 10 kΩ when illuminated. It is in series with a 10 kΩ resistance in a potential divider, the input to which is 5 V. What is the p.d. across the cell (a) in the dark, (b) when illuminated?

6 Two 100 Ω strain gauges on a beam are in series as a potential divider. The supply p.d. is 1 V. What is the current? When the beam bends, one gauge increases in resistance by 0.1 Ω and the other decreases by 0.1 Ω. What is the current now? By how much does the p.d. across each strain gauge change?

Answers 1. 22.5 mV **2.** 3 A; zero; 1.0 V **3.** 0.25 Ω **4.** 5 mV **5.** almost 5 V; 2.5 V **6.** 5 mA; 5 mA as before; +0.5 mV; −0.5 mV

YOU HAVE LEARNED

- About sensors for sound, light, temperature, chemical substances and for strain.

- That the emf of a source is equal to its open circuit potential difference.

- That the potential difference from a source, emf E and internal resistance $R_{internal}$, is $V = E - IR_{internal}$, or $V = IR_{load}$, where I is the current drawn from the source, and R_{load} is the external load resistance in the circuit.

- About using a potential divider to measure potential differences.

- That operational amplifiers are used in handling signals from sensors.

- That response time is an important quality of a sensor.

- That temperature changes affect most measurements.

potential difference, internal resistance, potential divider, thermocouple, photodiode, thermistor, strain gauge, electromotive force, photoresistor, load

2.5 Making good sensors

Here are some of the qualities a good sensor system ought to possess:
- high resolution
- appropriate output for a given input (sensitivity)
- rapid response time
- small unsystematic fluctuations ('random error') in results
- small systematic error

Linearity is also useful, but with the advent of look-up tables is not as important as it used to be.

Resolution

The resolution of a sensor is the smallest change it can detect in the quantity it is measuring. Often in a digital display, the least significant digit will fluctuate, indicating that changes of that magnitude are only just resolved. The resolution is related to the precision with which the measurement can be made.

Sensitivity

The sensitivity of a measuring system is the ratio of change of output to change of input. For example, the sensitivity of an oscilloscope display might be stated in millimetres per volt. Sometimes it is necessary to reduce the sensitivity so that the measuring instrument can deal with larger changes of input. A potential divider is often useful to scale down a potential difference by a known factor.

Response time

The response time is the time a sensor takes to respond to a change in input. Changes which occur more rapidly than this will usually be averaged out. In any monitoring of a process, the response time of the measuring system must be short enough to detect important changes as they occur.

Noise, random error, fluctuations

The input signal may fluctuate or the sensor itself may generate noise. Small unsystematic variations are present in all experimental data. Their size limits the precision with which a measurement can be made. Taking an average over repeated measurements can improve the final result, as long as the conditions can be kept the same.

Systematic error

Systematic error is very hard to detect, because detecting it means making another, even better, measurement. Systematic errors include zero error, and error due to disturbing influences, for example temperature.

Today, the trend is towards 'smart' systems, which process information to compensate for disturbing influences. For example, a 'smart' system might simply measure the temperature and adjust the results automatically. Again, cheap software can improve the performance of inexpensive hardware. Hardware costs are cut.

Handling data

Data from a sensor system must be processed, often with appropriate averaging, and displayed effectively to communicate a message.

Creative invention: sensing position and displacement

We want to conclude this chapter by emphasising the inventive, creative aspect of making sensors. Anyone can join the game. Any physical principle or effect can be pressed into service. Some are well understood, some are not. Those that are not have to be checked out experimentally.

The panel on page 55 shows a wide variety of ways of sensing how near an object is, or how much it moves, or even just its presence. They cover a wide range of scales, from nanometres to kilometres. Some we have already mentioned. Others are new, and some use physics which comes later in the A level course—but no matter. Look at them for interest, and see if you can add an idea to the collection. As a part of this course, you may make a sensor of your own.

How to... Handle data:
- Fitting lines to data
- Dealing with variability in data
- Making charts and graphs
- Presenting data in tables

Nine ways to sense position, displacement, proximity

Potential divider

Principle:
Moving contact on a resistor taps off a p.d. proportional to displacement.

Remarks:
Cheap, simple, limited range.

Radar

Principle:
Radio pulse reflected from distant object. Distance from time taken by echo.

Remarks:
Long range, can have high precision.

Ultrasound

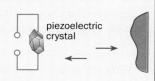

piezoelectric crystal

Principle:
Ultrasonic pulse reflected by nearby material. Distance from time taken by echo.

Remarks:
Use in medicine; also camera range-finders.

Reflected light

light emitting diode

light detecting diode

Principle:
Light from a diode is scattered back to a detector.

Remarks:
Cheap, simple, limited range and resolution.

Infrared detection

pyroelectric detector

Principle:
A charge on the crystal depends on temperature. Infrared radiation warms the crystal, changing the charge.

Remarks:
Used in burglar alarms.

Moiré fringes

Principle:
Two grids of lines at an angle form 'fringes'. Moving one grid moves the fringes, detected by a photocell.

Remarks:
Can detect down to 1 µm.

Differential transformer

primary coil

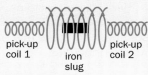

pick-up coil 1   pick-up coil 2
iron slug

Principle:
If the slug moves from dead centre the pick-up coils have unequal alternating p.d.s.

Remarks:
Capable of high resolution.

Capacitance

Principle:
The charge on a plate near a conducting surface depends on the spacing.

Remarks:
Works well on µm scale. Non-linear.

Scanning probe

Principle:
A fine tip near a surface collects an electron tunnelling current.

Remarks:
Can resolve atoms and molecules.

Try these

1 One junction of a thermocouple gives an output of 50 μV per degree Celsius. If the detecting instrument can just detect a change of 0.1 mV in p.d., what change in temperature can just be resolved? What is the resolution if ten such junctions are used in series in a thermopile?

2 Estimate the smallest change in amount of petrol in the tank which a typical car fuel gauge can show.

3 Estimate the resolution of the skin of your hand, as a temperature detecting device.

4 The mains frequency is 50 Hz, taking 20 ms to complete one cycle. Set a limit on the response time of a photosensor which could detect the rise or fall of intensity of light from a mains lamp.

5 Some digital stop-watches can indicate down to 1/100 s. Estimate the possible systematic error in timing a race, due to the reaction time of the person using the stop-watch. Think of other sources of systematic error in timing a race.

6 Suggest two different ways to a make a sensor which produces an electrical signal that indicates the temperature of its surroundings. Compare their likely response times. Would either method be likely to require the use of (a) a potential divider, (b) an amplifier?

7 Discuss the qualities of a potential divider as a position or angle sensor, under each of the headings: resolution, sensitivity, response time, noise, systematic error (see page 54).

Answers 1. 2°C, 0.2°C **4.** less than 5 ms

YOU HAVE LEARNED

- About qualities of a sensor system:
  - resolution
  - sensitivity
  - response time
  - random error
  - systematic error
- That the design of sensors can make use of a wide variety of physical principles

resolution, sensitivity, response time, random error, systematic error

Summary checkup

Electricity:

- Electric current = charge transferred/time taken; $I = \Delta Q / \Delta t$

- Potential difference = potential energy difference/charge transferred; $V = \Delta E / \Delta Q$

- Conductance $G = I/V$; Resistance $R = V/I$ (neither necessarily constant).

- A conductor which obeys Ohm's law has a constant conductance and resistance, so that the current is directly proportional to the potential difference.

- Resistances in series add up: $R = R_1 + R_2$

- Conductances in parallel add up: $G = G_1 + G_2$, and hence $1/R = 1/R_1 + 1/R_2$

- The potential difference from a source, emf E and internal resistance $R_{internal}$, is $V = E - IR_{internal}$, or $V = IR_{load}$, where I is the current drawn from the source, and R_{load} is the external load resistance in the circuit.

- Power in a circuit $P = IV = V^2/R = I^2 R$

Sensors:

- Sensors can be made on a scale of micrometres, often in silicon.

- Self-generating sensors include thermocouples, photovoltaic cells and piezoelectric crystals.

- Modulating sensors include potential dividers, photoconductive cells, thermistors and other resistances which change with temperature, gas-sensitive materials and strain gauges.

- Important qualities of a sensor are:

    Good resolution
    Appropriate sensitivity
    Rapid response time
    Small unsystematic random errors, perhaps reduced by averaging
    Small systematic or zero error

- Sensor system design may require:

    Dealing with lack of linearity, perhaps using look-up tables
    Preventing results from being affected by other changes, notably temperature

Circuits:

- Potential dividers can produce a p.d. as output from a change in resistance of a sensor.

- Operational amplifiers are used in handling signals from sensors.

- Circuits based on the potential divider have many uses.

Questions

1 **Paper uniformly coated with graphite can be electrically conducting. A strip of the coated paper 10 mm wide and 250 mm long has a resistance of 2 kΩ.**

(a) What is the conductance of the strip, expressed in mA V^{-1}?

(b) What would be the resistance of a strip twice as wide and twice as long?

(c) What would be the conductance of a strip 40 mm wide and 250 mm long?

(d) What would be the resistance of a strip 20 mm wide and 1 m long?

(e) Suggest dimensions for a strip having resistance 500 Ω.

(f) Suggest dimensions for a strip with conductance 1 mA V^{-1}.

2 **Give and explain an example of a potential divider used to produce an electrical output signal.**

(a) Include in your explanation:

● *a labelled circuit diagram showing where the output signal is obtained and where an input p.d. needs to be applied*

● *what determines the magnitude of the output signal*

● *what quantity the output signal indicates, and how it does so*

● *the effect on the output signal of attempting to measure it with a low resistance voltmeter.*

(b) Describe a practical application for a potential divider used in the way you describe. State one difficulty which would have to be overcome in obtaining accurate results in this application.

3 **Electrons are accelerated through a p.d. of 10 kV, and focused into a beam which falls on a circular spot 1 mm in diameter on a screen. The beam current is 2 mA.**

Calculate:

(a) the number of electrons falling each second on the spot

(b) the energy of one electron in the beam, in joules

(c) the power in watts delivered to the spot.

Estimate:

(d) the number of electrons per second striking one atom in the area of the spot.

4 **The graph shows how the resistance of a thermistor varies with temperature.**

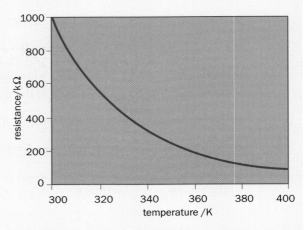

(a) Write a caption for the graph, which says what it shows about how the resistance varies with temperature.

(b) Sketch a graph of how the conductance varies with temperature over the same range. Explain the shape of your graph.

(c) If the thermistor is connected in series with a 1 kΩ fixed resistor to a source of emf 1.5 V with negligible internal resistance,

(i) at what temperature would you expect the p.d. across the thermistor to be equal to that across the fixed resistor?

(ii) at a certain temperature the p.d. across the thermistor is 0.5 V. What is its resistance at this temperature? Estimate the temperature.

(iii) approximately what p.d. do you expect across the thermistor at a temperature of 400 K?

(d) If the source had appreciable internal resistance, would the effects of this on the measurement be more important at high or at low temperatures? Why?

5 **Give and explain an example of each of the following:**

(a) a sensor which has insufficient resolution

(b) a sensor which has too large a response time

(c) a sensor or measurement which has an important systematic error.

In each case, suggest a possible way of dealing with the problem.

3 Signalling

Signals annihilate distance, bringing people far apart into contact. In this chapter we tell the story of the digital revolution in communication, a revolution with consequences not yet foreseen. We will give examples of:

- digital transmission of information

- uses of the electromagnetic spectrum in signalling

- calculations of limits on signalling capacity

3.1 Digital revolution and the death of distance

About the end of the first century AD, a group of Dacians crossed the river Danube from their home-land—now Romania—and attacked the army of the Roman Emperor Trajan. The Romans lit signal fires to call for help. The fires were repeated from camp to camp, reserves arrived, and the Dacians were defeated. Beacons communicated news for more than a thousand years. Today, satellites watch the whole surface of the Earth.

By 1793 the young experimenter Claude Chappe had persuaded leaders of the French Revolution of the military and political advantages of a semaphore telegraph system. By 1798 Government messages could travel the 400 km between Paris and Strasbourg in 36 minutes. 'By this invention distances between places disappear...it is a means which tends to consolidate the unity of the Republic', declared a member of the Committee of Public Safety. By contrast, a hundred years earlier Robert Hooke had seen semaphore as a means of private communication, as 'a method for making your thoughts known far away'.

• Semaphore unifies the new republic

Engraving of a telegraph station inaugurated at Condé in France in November 1794. Operators read signals from another station by telescope, and sent them on in the same way to the next. Earlier, in 1791, when Chappe built a demonstration semaphore tower in Paris, the townsfolk tore it down during the night.

The original telegraph transmitting (bottom of picture) and receiving (top of picture) devices, as designed by Samuel Morse. Instead of an on–off switch worked by the operator, the transmitter worked by having the Morse code message set up in metal type and pulled through the device, with an electrical contact to the metal type turning the signal into electrical impulses. At the receiver, incoming pulses pulled down a pen which wrote the message on a moving paper tape.

Hooke was right, as well as Chappe. First the telephone, then fax, electronic mail and mobile phones, have transformed personal communication. As families living in the same place broke up, the telephone helped keep smaller dispersed family units in touch. The Internet, first devised for the US military, has become a worldwide personal means of communication.

Commercial interests caught on. Bankers used the semaphore telegraph systems to get news of the latest prices (in 1836 two bankers in Bordeaux were caught secretly adding such information to Government messages). Today, stock market prices all round the world appear on dealers' screens.

When in 1876 Alexander Graham Bell said 'Mr Watson, come here, I want to see you' he was speaking into one of the only two telephones in the whole world—Elisha Gray had applied for a patent on the same day. Two lifetimes later there is one telephone to every five or six human beings on Earth.

The transatlantic telephone cable completed in 1956 could carry just 89 simultaneous conversations, serving all of Europe communicating with all of North America. By 1996, the total capacity was 1.3 million lines, following the laying of fibre optic cables. Transatlantic telephone links by satellite added a further 710 000 lines by the same date. International calls, once scarce and expensive, are now frequent and cheap.

On an April afternoon in the year 1912 David Sarnoff made radio history. The young Marconi telegraph operator picked up faint Morse code messages from the sinking ship *Titanic*, and became the one link the stricken vessel had with the outside world. In 1916 the Marconi Company rejected his proposal for broadcasting music to people at home. But within five years, public radio broadcasting began, using his ideas. Today, telecommunications is a worldwide business, with Asian countries in the forefront. The percentage of homes with television reaches nearly 100% in many parts of the world. A fact of political importance is that it is already over 60% in China. Digital and cable television channels have exploded in number.

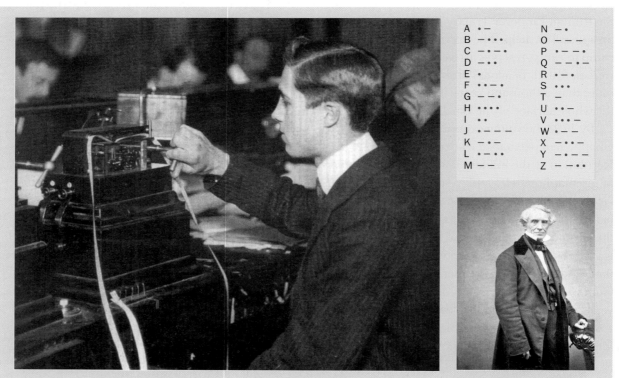

A telegraph operator in 1909, receiving a Morse code message from India. Samuel Morse (1791–1872), right, was the first to have the idea of a single-wire telegraph communicating characters by means of a code. His code was designed so that the most frequently used letters had the shortest codes. He was also well known as a portrait painter. For many years Morse ('wireless telegraphy') was the sole means of rapid world-wide communication. Telegraph operators built up skill in sending Morse code, and, more important, hearing and decoding it. New technologies create new jobs and destroy old ones.

How will all this change people? Is it good for democracy that your vote may be swung by how well politicians have learned to perform on television? Does viewing tragic world events and being helpless to do anything about them affect people's sense of responsibility? Could being able to contact almost anyone at almost any time you want change human communities?

Going digital: fax and electronic mail

The Roman signal fire was a one-bit digital signal: 'on' meant 'send help'. Like other digital signals its virtue was that it could easily be relayed—by lighting the next fire in the chain.

Alexander Graham Bell's telephone was not digital. The vibrations of his voice were changed into matching oscillations of an electrical potential difference, which travelled along a wire and were then changed back into sound vibrations in the air. This is analogue signalling. These telephones only worked if the signal was amplified at 'repeater' stations along the line, because the electrical oscillations got weaker as they travelled. Unfortunately the amplifiers boosted the level of any noise as much as they boosted the signal itself, and needed careful design to avoid distortion. Going digital has changed all that, because it is easier to detect 'on' or 'off' signals even when weakened and noisy, and codes which correct errors can be used. A perfect copy can be regenerated and sent on, new 'signal fires' lit.

Fax (facsimile) and e-mail (electronic mail) both send digital signals. A fax machine scans a page, treating it as a pattern of dots or pixels (see chapter 1). Each pixel is coded 1 or 0, according to whether it is nearest to being black or white, with a resolution of 200 pixels per inch—sometimes up-to-date technology uses ancient units!

To send a page, you need to send around 4 million bits, one bit per pixel. A commonly used rate of transmission on digital telephone lines is 64 000 bits per second (see page 65), so the job can take about a minute. Smart faxes speed it up by compressing the signal, sending the number of black or white bits to come in a long run of one kind, instead of sending the bits themselves. It's a lot quicker to send the binary

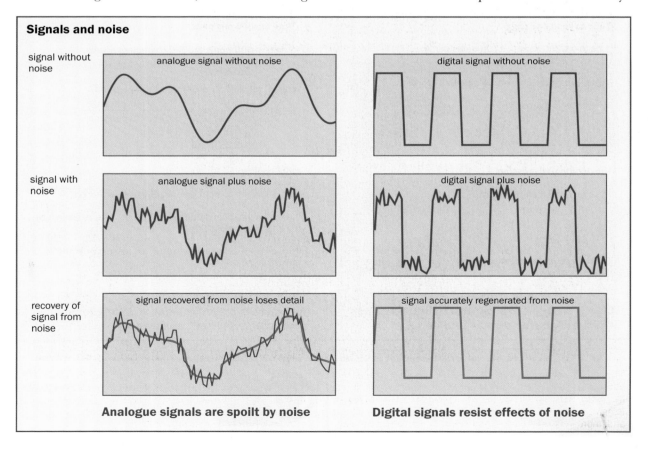

Signals and noise

signal without noise — analogue signal without noise | digital signal without noise

signal with noise — analogue signal plus noise | digital signal plus noise

recovery of signal from noise — signal recovered from noise loses detail | signal accurately regenerated from noise

Analogue signals are spoilt by noise   **Digital signals resist effects of noise**

digits of the number '1000' than to send a thousand bits. Digital television couldn't work without compression (page 73).

In electronic mail, each character is sent as a one-byte number code (ASCII). The 256 alternatives allow enough possibilities to check errors as well as to distinguish letters of the alphabet in capital and lower case, punctuation marks, numerals etc. A page taking several minutes to read aloud takes less than half a second to send.

E-mail is popular in countries which use the Roman alphabet, except for documents which require a valid signature or graphic logo. In countries like Japan, where the written language uses thousands of pictorial characters ('kanji'), fax is easier.

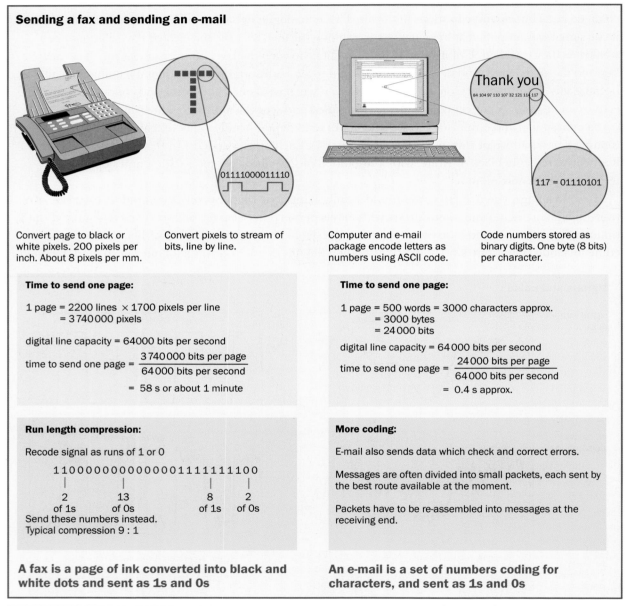

Sending a fax and sending an e-mail

Convert page to black or white pixels. 200 pixels per inch. About 8 pixels per mm.

Convert pixels to stream of bits, line by line.

`01111000011110`

Computer and e-mail package encode letters as numbers using ASCII code.

Thank you
84 104 97 110 107 32 121 114 117

Code numbers stored as binary digits. One byte (8 bits) per character.

`117 = 01110101`

Time to send one page:

1 page = 2200 lines × 1700 pixels per line
      = 3 740 000 pixels

digital line capacity = 64 000 bits per second

time to send one page = $\dfrac{3\,740\,000 \text{ bits per page}}{64\,000 \text{ bits per second}}$

      = 58 s or about 1 minute

Time to send one page:

1 page = 500 words = 3000 characters approx.
      = 3000 bytes
      = 24 000 bits

digital line capacity = 64 000 bits per second

time to send one page = $\dfrac{24\,000 \text{ bits per page}}{64\,000 \text{ bits per second}}$

      = 0.4 s approx.

Run length compression:

Recode signal as runs of 1 or 0

`1 1 0 0 0 0 0 0 0 0 0 0 0 0 0 0 0 1 1 1 1 1 1 1 1 0 0`

2   13   8   2
of 1s   of 0s   of 1s of 0s
Send these numbers instead.
Typical compression 9 : 1

More coding:

E-mail also sends data which check and correct errors.

Messages are often divided into small packets, each sent by the best route available at the moment.

Packets have to be re-assembled into messages at the receiving end.

A fax is a page of ink converted into black and white dots and sent as 1s and 0s

An e-mail is a set of numbers coding for characters, and sent as 1s and 0s

● Data on the telecommunications explosion

● A poem about messages

Digital music

You slip a shiny compact disc into the player, it spins, and out of your loudspeakers comes top quality stereo sound, with no audible hiss or noise. A little dust, even a scratch, on the disc may be ignored—or rather, your CD player works out what it should have found and plays that to you. This is what digital signal processing can do.

A scanning electron microscope shows the tiny marks which encode the music as a stream of bits: shallow pits cut into its upper surface circle the disc in a long spiral track. The spiral has only 1.6 μm between turns. Lenses focus light from a laser diode into a tiny spot on the surface. The light is reflected back to a light-sensitive sensor which converts the light signal to an electrical signal.

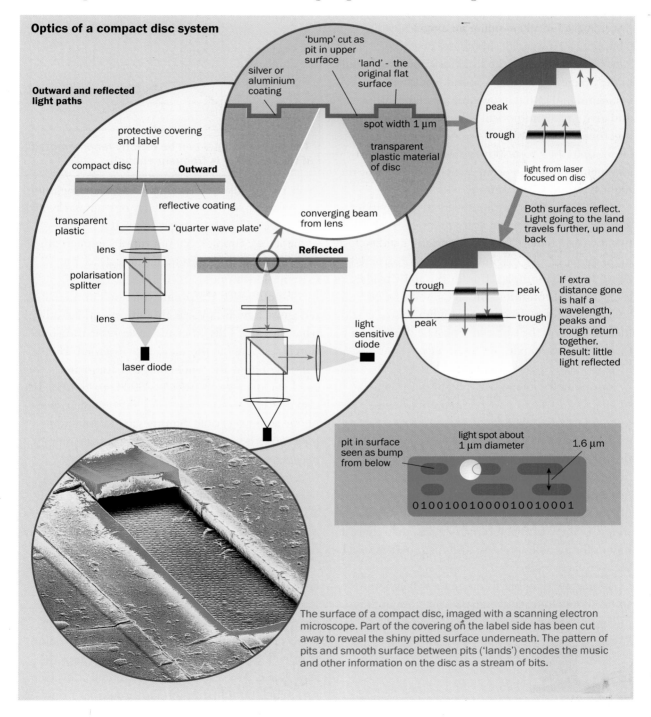

Optics of a compact disc system

Outward and reflected light paths

protective covering and label

compact disc

Outward

transparent plastic

reflective coating

'quarter wave plate'

lens

polarisation splitter

lens

laser diode

'bump' cut as pit in upper surface

silver or aluminium coating

'land' - the original flat surface

spot width 1 μm

transparent plastic material of disc

converging beam from lens

Reflected

light sensitive diode

peak

trough

light from laser focused on disc

Both surfaces reflect. Light going to the land travels further, up and back

trough

peak

peak

trough

If extra distance gone is half a wavelength, peaks and trough return together. Result: little light reflected

pit in surface seen as bump from below

light spot about 1 μm diameter

1.6 μm

0100100100001001 0001

The surface of a compact disc, imaged with a scanning electron microscope. Part of the covering on the label side has been cut away to reveal the shiny pitted surface underneath. The pattern of pits and smooth surface between pits ('lands') encodes the music and other information on the disc as a stream of bits.

Digitising sounds

The telephone is another example of signals now usually transmitted digitally. Digitising sound is done in three steps:

- sampling
- binary coding
- further encoding

Sampling

To digitise an image, you divide it up into tiny discrete pixels and give each a numerical value (chapter 1). How small should the pixels be? Answer: smaller than the smallest important detail in the image. To digitise a sound, you divide the varying signal into narrow time slices, or samples, and give each a numerical value. How close together in time should the samples be? Answer: closer than the shortest time in which important changes in the signal occur, which is decided by the highest frequency it contains.

In this way, the continuous signal is turned into a string of samples taken at frequent intervals. You may be surprised to learn that nothing need be lost in the process. The original can be reconstructed *exactly* from the samples. But for this to be true there is an essential condition: the signal must be guaranteed to contain no frequencies higher than a certain maximum, and the sampling must be done faster than twice that maximum frequency.

sampling frequency > 2 × highest frequency component present

If there is a limit to the high frequencies present, the waveform cannot contain any unexpected 'wiggles' between closely spaced samples. The original signal can be recovered exactly by joining up the samples with smooth curves of the right shape. If there are such rapid wiggles, then there *are* higher frequencies present, and the signal can *not* be recovered correctly.

Worse, such higher frequencies will generate spurious low frequency signals which are not there

● Simple sampling

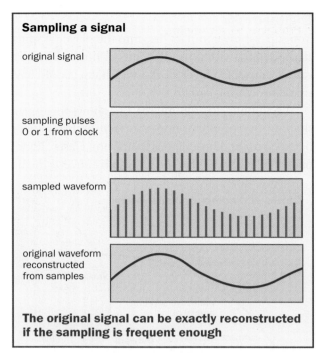

Sampling a signal

original signal

sampling pulses
0 or 1 from clock

sampled waveform

original waveform
reconstructed
from samples

**The original signal can be exactly reconstructed
if the sampling is frequent enough**

at all, called 'aliases'. You have seen the same thing when wheels seem to turn slowly or even backwards in a movie. The film images come as a rapid sequence of samples, but not rapidly enough, and very fast repeating movements appear as non-existent slower ones.

Sometimes on the phone you don't immediately recognise a friend's voice, though you understand what they say. This is because telephone transmission is restricted to the small slice from 300 Hz to 3400 Hz out of the whole range of frequencies you can hear, perhaps from below 50 Hz to 20 kHz. It's also the reason why music played from an answerphone sounds terrible. Samples are taken about 8000 times a second, a little faster than absolutely necessary.

On a music CD frequencies up to 20 kHz are needed, so the sampling must be done more than 40 000 times a second. 44.1 kHz is standard. Filters remove frequencies above 20 kHz, which would give trouble.

Binary coding

Look back to the antenatal scan image on page 1, chapter 1. Each pixel in it is assigned one of 256 levels of grey, using 8 bits or 1 byte of information per pixel. Sound sampling for telephony uses the same scale. An 8 bit 'analogue to digital converter' turns each sample strength into a number from 0 to

Problems with sampling

Sampling too slowly misses high frequency detail in the original signal

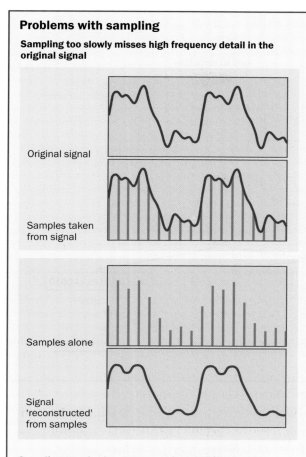

Original signal

Samples taken from signal

Samples alone

Signal 'reconstructed' from samples

Sampling too slowly creates spurious low frequencies (aliases)

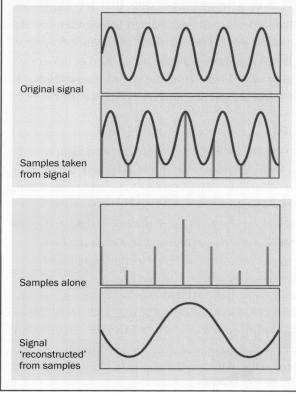

Original signal

Samples taken from signal

Samples alone

Signal 'reconstructed' from samples

255, on the binary scale. This is quite coarse, but good enough for intelligible speech. Higher quality CD sound requires better resolution: it uses 16 bits, giving 65 536 different levels for each sample.

The telephone signal has now been converted into a stream of bits, a 'bit stream'. They come in blocks of eight, each block giving the magnitude of one sample. Obviously, the eight bits for one sample must be sent before it is time for the next sample to come along 1/8000 of a second later. So the bits must be sent at a rate of at least 64 000 bits per second.

CD quality sound demands a much higher channel capacity. 16 bit samples must be sent at a rate of 44.1 kHz, so the bits must come along 16 times faster, at 0.7 MHz. Squeezing in two stereo channels doubles the rate to 1.4 MHz, a frequency similar to those in the medium radio wave band (see page 68).

In general, then, the channel capacity needed to transmit a digitally sampled signal depends on:

- the resolution of each sample, that is the number of bits specifying the value of each sample
- the rate of sampling, which must be at least twice the highest frequency in the signal spectrum.

Digitising a signal

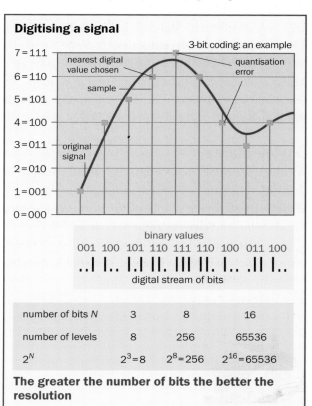

$7 = 111$
$6 = 110$
$5 = 101$
$4 = 100$
$3 = 011$
$2 = 010$
$1 = 001$
$0 = 000$

3-bit coding: an example

nearest digital value chosen

quantisation error

sample

original signal

binary values
001 100 101 110 111 110 100 011 100

digital stream of bits

number of bits N	3	8	16
number of levels	8	256	65536
2^N	$2^3 = 8$	$2^8 = 256$	$2^{16} = 65536$

The greater the number of bits the better the resolution

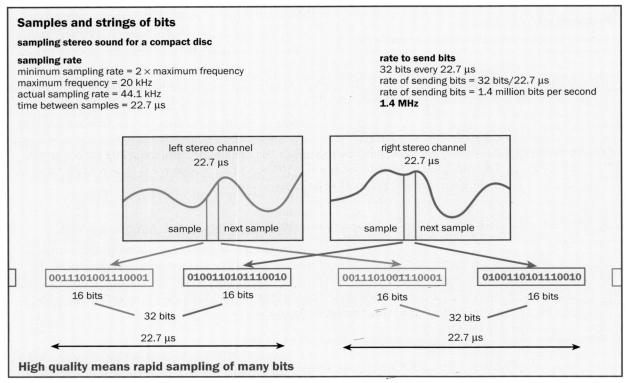

Samples and strings of bits

sampling stereo sound for a compact disc

sampling rate
minimum sampling rate = 2 × maximum frequency
maximum frequency = 20 kHz
actual sampling rate = 44.1 kHz
time between samples = 22.7 μs

rate to send bits
32 bits every 22.7 μs
rate of sending bits = 32 bits/22.7 μs
rate of sending bits = 1.4 million bits per second
1.4 MHz

left stereo channel
22.7 μs

sample   next sample

right stereo channel
22.7 μs

sample   next sample

0011101001110001   0100110101110010   0011101001110001   0100110101110010

16 bits   16 bits   16 bits   16 bits

32 bits   32 bits

22.7 μs   22.7 μs

High quality means rapid sampling of many bits

Further encoding

People take it for granted that electronic mail will arrive without errors, and expect CD sound to be fault-free. In fact, any transmission of a signal introduces errors; a few bits get lost or changed because of noise. But in spite of that, digital signal transmissions are quite remarkably free of errors. This is achieved by coding the signals in such a way that errors can be located and corrected. Actual error-correction codes are very complicated, but depend on a simple basic idea—sending more information than the minimum. This is called adding redundancy. A crude way to do it would be to send each part of the signal more than once. If the 'copies' do not agree, there must be an error, and they can be asked for again until they agree. Much cleverer ways than this are used in practice, though making repeated readings is used in bar-code readers at supermarket tills. The essential point is that using digital signalling, the system *computes* what it should have sent. So tolerant is a CD that you can drill a 2 mm hole in one and barely notice the effect.

Digital futures

A consequence of the fact that digital information is just numbers which can be used in a computation is that different kinds of information can be mixed: e-mails with images or even tunes attached, for example. Digital signals, being numbers, are also easy to encrypt or 'scramble'. This is going to be a big issue for the future: who will have the right to code and decode different signals?

As electronic commerce develops, will it be possible to keep your credit card number secure as it is sent over the Internet? What kinds of material will you normally have delivered to you as electronic files: newspapers? books? music? brochures? videos? pictures to put on your wall? Can ownership of such information be protected? Should it be? The digital future poses many such problems.

● Digital recording error correction

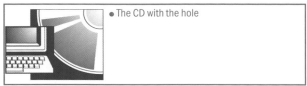
● The CD with the hole

Try these

1 If you chat on a digital telephone for half an hour, how many bits of information get transmitted, if 8 bit samples are sent 8000 times a second?

2 Try reading a paragraph of this book aloud, timing yourself. You can probably read about 200 words a minute. At an average of 6 characters per word, how long would 200 words take to send by e-mail, with 1 byte per character and a telephone channel sending 64 000 bits per second?

3 What 8 bit binary code is sent by e-mail when a space, with ASCII code 32, is to be sent?

4 How many bits must be read to get one minute of high quality stereo music from a CD, maximum frequency 20 kHz, with 16 bit sampling? How many 1 megabyte floppy discs would it fill?

5 The recording area of a CD measures 35 mm from inside to outside, and has an average radius of 40 mm. The recording track spirals with a spacing of 1.6 μm. Estimate how many turns the spiral makes. Estimate the length of the spiral. If the track is scanned at 1.25 m s^{-1}, estimate the playing time.

Answers 1. 115 million **2.** less than 0.2 s **3.** 00100000 **4.** nearly 80 million; about 10 **5.** 22 000 turns; 5.5 km; over one hour

YOU HAVE LEARNED

- The world has seen an explosive growth of signalling capacity.

- A signal channel has a capacity, the rate at which it can transmit information, measured in bits per second.

- Signals are increasingly digital; they can also be analogue. Digital signals have the advantage that:
  - they can be regenerated easily, reducing the effects of noise
  - they can be processed and encoded
  - they can represent different kinds of information in the same way.

- An analogue signal is made digital by sampling. The sampling must be done fast enough to reproduce the highest important frequencies in the signal.

- Samples can be read with different resolution (e.g. 8 bit, 16 bit). N bits gives 2^N different levels of measured values.

- Special coding methods can make digital transmission essentially error-free.

bit, amount of information, digital sampling, analogue to digital converter, aliasing

3.2 Signalling with electromagnetic waves

Spin the tuning dial of a radio (or press the 'seek' button) and you hear, one after the other, stations each at a particular frequency. The radio sits in a space filled with signals from all of them, all on top of one another. To hear one station, the radio must be 'tuned' to select just that one frequency. You could map out all the radio frequencies on a scale, with a line at each frequency, making it longer the stronger the signal. That would be a spectrum of the radio frequencies your set can pick up.

Imagine going for a swim in the sea. Long high waves lift your body and let it drop. Shorter waves splash against your face. Tiny ripples pass unnoticed. You are swimming in a spectrum of water waves, all just movements of the water surface happening at once. Now imagine standing outside in the sun. Rapid electromagnetic oscillations at infrared frequencies fall on your skin and warm you. Electromagnetic waves you don't feel pass around you: you know they are there when you turn the radio on. A mobile telephone rings: more radio waves are sent and received. You are swimming in a sea of electromagnetic waves, all just oscillations of electric and magnetic fields in empty space, all on top of one another.

Communication wave bands

frequency			wavelength
30 kHz			10 km
	LF low frequency	navigation, radio beacons, long distance broadcasting	
300 kHz			1 km
	MF medium frequency	national broadcasting, aeronautical navigation	
3 MHz			100 m
	HF high frequency	long distance broadcasting, amateur radio, maritime radio	
30 MHz			10 m
	VHF very high frequency	FM radio, mobile radio communications	
300 MHz			1 m
	UHF ultra high frequency	television, mobile telephone networks	
3 GHz			100 mm
	SHF super high frequency	satellite links, ground microwave links, radar	
30 GHz			10 mm
	EHF extremely high frequency	radar, radio astronomy	
300 GHz			1 mm
	far infrared		
3 THz			100 μm
	mid infrared		
30 THz			10 μm
	near infrared	optical fibre, remote controls, bar codes, CD player	
300 THz			1 μm

Swimming in a sea of electromagnetic waves

infrared from sun

mobile phones

light reflected from water

transistor radio

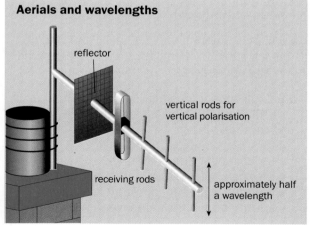

Aerials and wavelengths

reflector

vertical rods for vertical polarisation

receiving rods

approximately half a wavelength

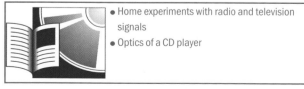

- Home experiments with radio and television signals
- Optics of a CD player

When radio broadcasting in Britain first began in the early 1920s, a single 1500 m wavelength signal, frequency 200 kHz, covered most of the country, its long wavelength meaning that hills hardly got in the way (see chapter 6). Today, frequencies used for communication stretch up to those of the infrared waves used in optical fibres. In between are the microwaves used for telephone links overland and to satellites.

Looking at the radio and television aerials on the roof of a house can give you an idea of the wavelengths being used. Typically, the aerial rods are half a wavelength long. Television uses the ultra high frequency UHF band 300 MHz–3 GHz, as do mobile telephones. The wavelength is a few tenths of a metre. FM radio uses the VHF band from 30 to 300 MHz, with wavelengths of a few metres. Satellite transmissions use microwaves with wavelengths of a few tens of millimetres; the aerial is now a collecting horn at the focus of a dish.

Polarisation

You can tell something else about a signal by looking at an aerial on the roof: its direction of polarisation. Radio waves arriving make electrons surge up and down the aerial rods at the frequency of transmission. The transmitter sent the radio waves by making currents oscillate in a similar rod. The varying electric field in the wave points parallel to the rods, parallel to the direction of motion of the electrons. To pick up the signal, the receiving aerial rods must be parallel to the changing electric field.

Generally radio signals are either vertically or horizontally polarised. It happens that on my roof, the FM radio aerial is aligned for horizontal polarisation, and the TV aerial for vertical polarisation.

A sound wave cannot be polarised like this, because the oscillations in the wave are along the direction of wave propagation: the wave is longitudinal. A radio wave can be polarised, because the oscillating electric and magnetic fields are at right angles to the direction of propagation: the wave is transverse.

If you go fishing or skiing you may wear Polaroid spectacles. They cut out visible light which is polarised in one direction, as light reflected from water is. Light scattered high in the atmosphere, making the sky blue, is also polarised, and some insects use this direction of polarisation to navigate their way from nest to plant.

The CD system also makes clever use of polarised light. Light is steered through the system by a 'polarising prism' which transmits light polarised in one direction and reflects light polarised in the other direction.

What's in a signal?

Press two notes on a piano or pluck two strings on a guitar. Your ear takes the complex resulting vibration and analyses it into two notes—the spectrum of the sound. This is why musicians can learn to identify the notes in a chord. Your eye does not do the same for light, however. A mixture of red and blue looks purple, not 'red plus blue'. Computer programs can take a repeating signal and analyse it to see what frequencies

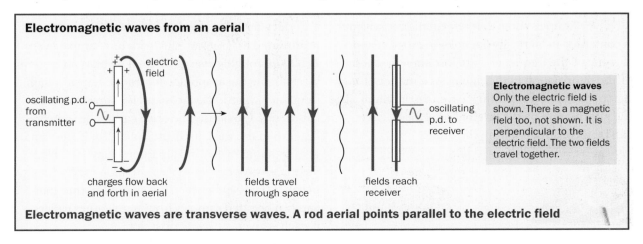

Electromagnetic waves from an aerial

oscillating p.d. from transmitter

electric field

charges flow back and forth in aerial

fields travel through space

fields reach receiver

oscillating p.d. to receiver

Electromagnetic waves
Only the electric field is shown. There is a magnetic field too, not shown. It is perpendicular to the electric field. The two fields travel together.

Electromagnetic waves are transverse waves. A rod aerial points parallel to the electric field

Polarisation by scattering

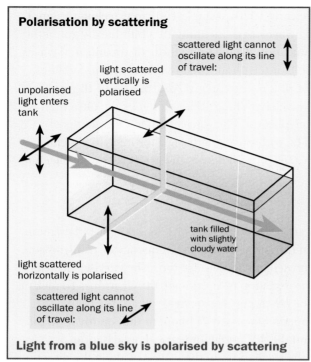

scattered light cannot oscillate along its line of travel:

light scattered vertically is polarised

unpolarised light enters tank

tank filled with slightly cloudy water

light scattered horizontally is polarised

scattered light cannot oscillate along its line of travel:

Light from a blue sky is polarised by scattering

Waveforms and spectra of sounds

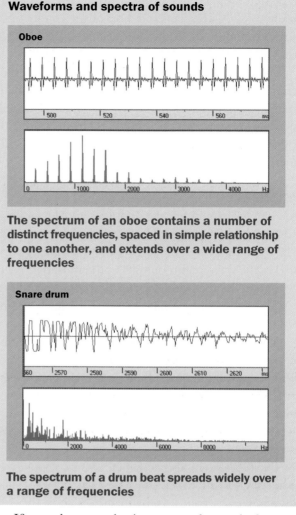

Oboe

Snare drum

The spectrum of an oboe contains a number of distinct frequencies, spaced in simple relationship to one another, and extends over a wide range of frequencies

The spectrum of a drum beat spreads widely over a range of frequencies

it contains. Your ear can do it; your eye cannot. A prism does it for light, and a tunable radio set does it for radio waves. If the signals are digitised, the computer analysis can be done rather quickly: you can see the spectrum of a sound as you make it. The analysis (Fourier analysis) is like a mathematical tuning dial, picking out frequency components one by one.

Think for a moment about hearing impairment. One kind makes the ear insensitive to high frequencies (this happens also as you grow older). The high frequency components in sounds such as 's' and 't' become inaudible. 'Bit' may be hard to tell from 'bid'. Tapes of sounds filtered to be like those heard by people with various impairments can help you to understand their problems.

The examples show that information in a signal is contained in all the frequencies in its spectrum. If some of the frequencies are lost, then so is some of the information in the signal. As a matter of fact, the spectrum of a repeating waveform contains all the information of the original waveform. Each can be reconstructed from the other.

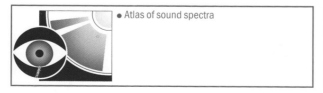

● Atlas of sound spectra

If you play a synthesiser you make musical tones by mixing the required frequencies: you go from spectrum to sound. When you identify a chord or use a computer program to analyse a sound you are going the other way, from sound to spectrum.

Sending many signals at once

You call a friend in another country. Your call is transmitted as 8 bit samples, with one sample every 1/8000 s, roughly every 100 µs. The bits can be sent by changing the amplitude (or frequency) of a high frequency wave backwards and forwards between two values. Obviously the wave cannot be switched faster than the wave itself oscillates. But if the signal goes by a microwave link, frequency say 10 GHz, the microwave carrier makes a million oscillations in the 100 µs gap between your samples. Into that gap can be fitted samples from many other callers, if each is given a private time slot.

Cellular telephones

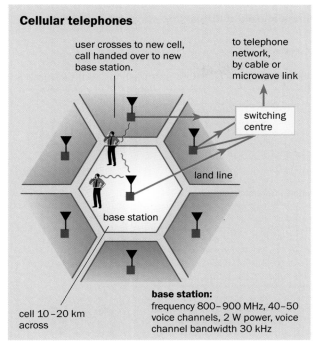

user crosses to new cell, call handed over to new base station.

to telephone network, by cable or microwave link

switching centre

land line

base station

cell 10–20 km across

base station:
frequency 800–900 MHz, 40–50 voice channels, 2 W power, voice channel bandwidth 30 kHz

Sending digital signals by radio

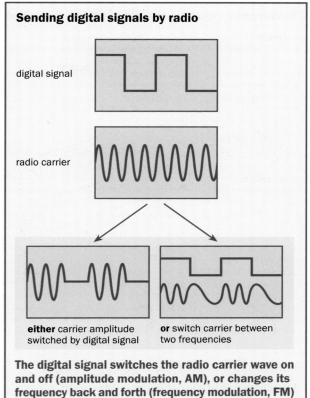

digital signal

radio carrier

either carrier amplitude switched by digital signal

or switch carrier between two frequencies

The digital signal switches the radio carrier wave on and off (amplitude modulation, AM), or changes its frequency back and forth (frequency modulation, FM)

You see how by using a high transmission frequency, many telephone channels can be fitted into one radio frequency band. Optical fibres, using much higher infrared frequencies, can carry many more bits per second; by 1990 it was 2 gigabits per second, room for 30 000 telephone channels.

Room on air

The width of a signal's spectrum—the range of frequencies it covers—is the **bandwidth** of the signal. Telephone conversations have a bandwidth of 3100 Hz (300 Hz to 3400 Hz). High fidelity music has a bandwidth of 20 kHz, or 40 kHz for stereo. The larger the bandwidth the faster bits must be sent.

You may have noticed how the frequencies of FM radio stations are spaced out. One may be 98.4 MHz; the next nearest will never be closer than 98.2 or 98.6 MHz, always at least 0.2 MHz or 200 kHz apart. This is because sending a signal on a radio carrier, whether digitally or not, spreads out the frequencies in the spectrum of the carrier.

Radio, television, mobile telephones, satellite and microwave links, and optical fibres, all use different parts of the electromagnetic spectrum (see page 68). The faster information needs to be transmitted, the larger the bandwidth and the higher the frequency band needed. The bandwidth required for a signal limits the number of stations which can

be fitted into a given waveband. Bandwidth is a valuable commodity.

Here's how to estimate the bandwidth required for a digital signal. Take the toughest job you can ask it to do, which is to send alternating bits 010101010101.... This is tough because the switching has to go at the fastest possible rate. It's a square wave. Its spectrum must have a fundamental sine wave component at the frequency of alternation of the repeating pairs of 0s and 1s. It has higher frequency components too, which 'sharpen up' the edges of the pulses. The very least the signal system must be able to do is to transmit and receive the frequency of these fast-repeating pairs. If it can't, then it can't respond quickly enough to transmit this 'worst case' signal. If it can do that, it can 'see' the bits, even though they now look like round-shouldered sinusoidal pulses. That is enough to be able to regenerate 'clean' new bits to send on. So the bandwidth must at least cover this frequency range.

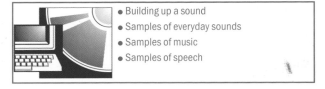

- Building up a sound
- Samples of everyday sounds
- Samples of music
- Samples of speech

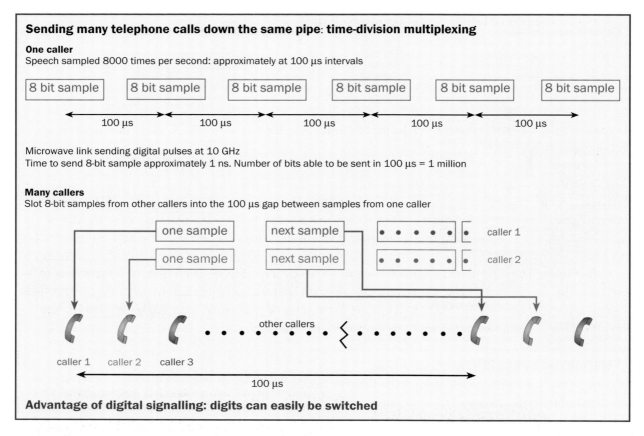

Sending many telephone calls down the same pipe: time-division multiplexing

One caller

Speech sampled 8000 times per second: approximately at 100 μs intervals

| 8 bit sample | 8 bit sample | 8 bit sample | 8 bit sample | 8 bit sample | 8 bit sample |

100 μs   100 μs   100 μs   100 μs   100 μs

Microwave link sending digital pulses at 10 GHz
Time to send 8-bit sample approximately 1 ns. Number of bits able to be sent in 100 μs = 1 million

Many callers

Slot 8-bit samples from other callers into the 100 μs gap between samples from one caller

one sample   next sample   • • • • • caller 1

one sample   next sample   • • • • • caller 2

other callers

caller 1   caller 2   caller 3

100 μs

Advantage of digital signalling: digits can easily be switched

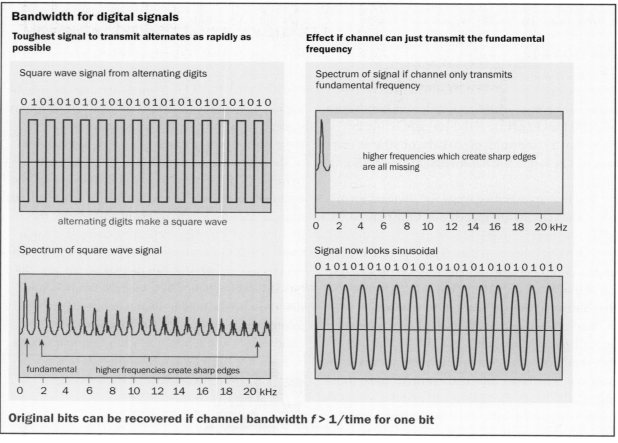

Bandwidth for digital signals

Toughest signal to transmit alternates as rapidly as possible

Square wave signal from alternating digits

0 1 0 1 0 1 0 1 0 1 0 1 0 1 0 1 0 1 0 1 0 1 0 1 0 1 0 1 0

alternating digits make a square wave

Spectrum of square wave signal

fundamental   higher frequencies create sharp edges

0   2   4   6   8   10   12   14   16   18   20 kHz

Effect if channel can just transmit the fundamental frequency

Spectrum of signal if channel only transmits fundamental frequency

higher frequencies which create sharp edges are all missing

0   2   4   6   8   10   12   14   16   18   20 kHz

Signal now looks sinusoidal

0 1 0 1 0 1 0 1 0 1 0 1 0 1 0 1 0 1 0 1 0 1 0 1 0 1 0 1 0

Original bits can be recovered if channel bandwidth $f > 1/$time for one bit

In this limit, a pair of bits '01' looks like one cycle of a wave.

maximum digital signalling rate in bits per second $= 2 \times$ bandwidth B needed

or

bandwidth $B = 1 / T$ where T is the time for which one bit is 'on'.

As a rough rule, take the bandwidth required to be of the order of magnitude of the maximum rate of transmission of bits.

Digital television

Television signals gobble up bandwidth, typically 8 MHz for each channel. Space in the UHF band is limited. But a satellite broadcasting at microwave frequencies up to 30 GHz may have a bandwidth of say 1 GHz, enough for over a hundred television channels. It is clear that the problem now is not providing more television channels, but is finding the time and money to put something worth watching on them all.

At first sight, digital television demands even more bandwidth. The huge number of bits needing to be sent has to be cut by compressing the signal. One method is to send after each frame only the pixels which change to give the next frame. Many frames have large areas of picture in common, so this can be very economical. If a frame doesn't change, nothing at all need be sent!

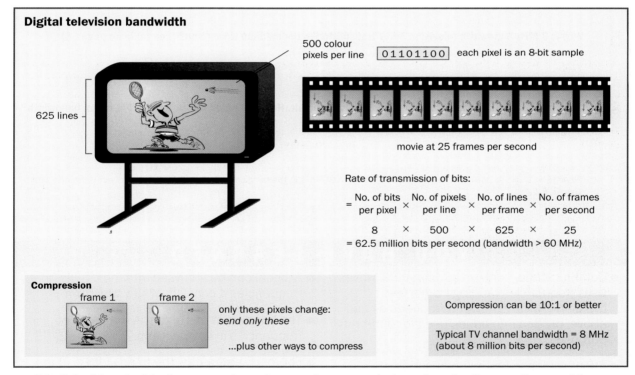

Digital television bandwidth

500 colour pixels per line   `01101100`   each pixel is an 8-bit sample

625 lines

movie at 25 frames per second

Rate of transmission of bits:

$$= \frac{\text{No. of bits}}{\text{per pixel}} \times \frac{\text{No. of pixels}}{\text{per line}} \times \frac{\text{No. of lines}}{\text{per frame}} \times \frac{\text{No. of frames}}{\text{per second}}$$

$$8 \times 500 \times 625 \times 25$$
$$= 62.5 \text{ million bits per second (bandwidth} > 60 \text{ MHz)}$$

Compression

frame 1   frame 2

only these pixels change: *send only these*

...plus other ways to compress

Compression can be 10:1 or better

Typical TV channel bandwidth = 8 MHz (about 8 million bits per second)

Of all the ideas in this chapter, perhaps the most important is the twin pair of a waveform and its spectrum—the picture in time and the picture in frequencies. This idea, under the name Fourier analysis, appears again and again in physics—in optics, in quantum theory and in many other places.

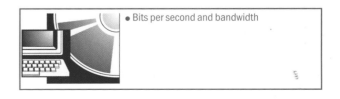

● Bits per second and bandwidth

Try these

1   A radio receiver has aerial rods each 75 mm long, making a pair half a wavelength long. What is its frequency?

2   The radio emission from some radio galaxies is polarised. How could that fact be detected?

3   What is the ratio of the highest to the lowest frequency in the table of radio bands on page 68 (excluding infrared)? What is the ratio of the longest to shortest wavelength?

4   Estimate roughly the bandwidth needed to send a 1000 word e-mail in one second, without compression.

5   Why is 8 MHz bandwidth television never transmitted in the medium wave band? (see table on page 68)

6   A fax page contains about 4 million pixels, each of one bit. How fast could one be sent by FM radio, bandwidth 200 kHz?

Answers 1. 1 GHz **3.** 10 million; also 10 million **4.** say 40 to 80 kHz, guessing between 5 and 10 characters per word including spaces **6.** 20 s

YOU HAVE LEARNED

- Communication with electromagnetic waves uses frequencies ranging from a few thousand hertz to infrared frequencies and above, divided into bands used for different purposes.

- About evidence for the polarisation of electromagnetic waves.

- A signal can be analysed into the frequencies it consists of—its spectrum.

- The bandwidth of a signal is the range of frequencies in its spectrum.

- For rough estimates, the bandwidth is of the order of magnitude of the maximum number of bits to be transmitted per second.

- Rates of transmission of digital information can be reduced by not sending information which is not necessary (compression).

polarisation, electromagnetic spectrum, radio waves, bandwidth, digital sampling

Summary checkup

✓ Signals

- Signals are digital or analogue. Digital signals are increasingly frequently used.

- Digital signals have the advantage that:

    they can be reproduced and re-transmitted easily
    they minimise the effects of noise
    they can be processed and encoded by computer
    they can represent different kinds of information in a uniform way.

✓ Sampling and encoding

- Analogue signals are digitised by sampling.

- Sampling must be done at a rate greater than twice the bandwidth (often, twice the highest frequency present).

- Samples can be read with different resolution (e.g. 8 bit, 16 bit). N bits gives 2^N different levels of measured values.

- Errors can be detected and removed if the signal is suitably encoded.

✓ Polarisation

- Electromagnetic waves, being transverse waves, can be polarised.

✓ Bandwidth, spectrum and rate of transmission of information

- Communication with electromagnetic waves uses frequencies from a few thousand hertz to infrared frequencies and above, divided into bands used for different purposes.

- A signal can be analysed into the frequencies it consists of—its spectrum.

- A signal channel has a capacity, the rate at which it can transmit information, measured in bits per second.

- The bandwidth of a signal is the range of frequencies in its spectrum.

- For rough estimates, the bandwidth is of the order of magnitude of the maximum number of bits transmitted per second.

Questions

Speed of electromagnetic waves $= 3 \times 10^8 \, m \, s^{-1}$

1 The spiral track of a music CD plays for one hour, read at 1.25 m s^{-1}. The two-channel stereo music on it has frequencies extending up to 20 kHz. The sampling frequency on each channel is 44.1 kHz. Samples are digitised with 16 bits per sample.

(a) Show that the track is 4.5 km long.

(b) Give a reason why the sampling frequency is more than 40 kHz.

(c) How many bits does the player have to read per second, for each stereo channel?

(d) Estimate the bandwidth of the communication between reading head and player.

(e) Show that there are 5×10^9 bits for 1 hour of stereo music.

(f) How much track length is there per bit? Comment.

2 The table below shows how the numbers of transatlantic and transpacific telephone lines have changed over time. Plot appropriate charts to display the changes and to compare cable and satellite. Suggest reasons for the changes in capacity which have occurred. (Consider whether to use a logarithmic scale or not.)

Data taken from Cairncross F *The Death of Distance* (1997) Orion Business Books. Sourced there to TeleGeography, Inc. (Washington, DC).

3 There is a photograph on the Web which I want to download. It is 30 kbytes in size. I do not want to wait more than half a minute for it to arrive. The picture is in colour, with 8 bits per pixel per colour.

(a) What rate of transmission in bits per second do I need the Web connection to provide?

(b) Can this rate be achieved down a telephone line of bandwidth 64 kHz?

(c) How many pixels should I expect the picture to have? If I print it 100 mm square, how big will a pixel be?

4 Cellular telephones use the frequency range 800–900 MHz. Speech frequencies up to 4 kHz are transmitted.

(a) What is the range of radio wavelengths used?

(b) At what rate must the speech be sampled?

(c) If samples are digitised with 1 byte per sample, how many distinct levels of signal can be sent?

(d) What rate of transmission in bits per second must the line achieve?

(e) Estimate the bandwidth used by one line.

(f) Estimate how many telephone lines can be handled in the 100 MHz band between 800 and 900 MHz.

5 Describe evidence for the polarisation of

(a) microwaves

(b) visible light

Suggest a practical application for the polarisation of one of them.

| Year | Transatlantic telephone lines | | Transpacific telephone lines | |
	Cable	Satellite	Cable	Satellite
1986	22 000	78 000	2 000	39 000
1987	22 000	78 000	38 000	39 000
1988	60 000	78 000	38 000	39 000
1989	145 000	93 000	38 000	39 000
1990	145 000	283 000	38 000	39 000
1991	221 000	283 000	114 000	27 000
1992	296 000	496 000	190 000	27 000
1993	410 000	621 000	264 000	83 000
1994	701 000	621 000	264 000	234 000
1995	1 311 000	711 000	264 000	234 000
1996	1 311 000	711 000	864 000	234 000
1997–2000 estimate	1 311 000	738 000	1 465 000	424 000

4 Testing materials

Here we begin the story of materials in use, their properties and how these properties depend on the underlying structure of the material. We will give examples of:

● different types of materials

● testing materials and properties

● uses of materials

The surgeon's sharp chromium-steel blade carefully slices into your eyeball. A fine tube sucks out the jelly-like contents of the lens. A clear polymer disc is inserted into the lens sac—and you can see again! This is now an everyday operation for people suffering from cataracts, but it wouldn't be possible without the materials and techniques which have been developed in the last few years.

Much of our way of life depends on modern materials—our communications and transport, health, sport and leisure, the progress of science itself. Most of the time we take this for granted, and just get on with our lives. But all the time people are designing materials to satisfy ever greater demands—for taller buildings, faster trains, smaller computers, longer-lasting body implants, a cleaner world.

This is a part of science where, every day, people are making new discoveries and producing new materials which have a direct impact on our lives.

Coalbrookdale by night, 1801. This remarkable oil painting by Philippe Jacques de Loutherbough (1740—1812) shows one of the Coalbrookdale ironworks at night silhouetted against the fiery glow of a furnace being tapped of its ore. The development of coke smelting in this area of Shropshire by Abraham Darby and his family in the 18th century revolutionised the production of iron and helped fuel the Industrial Revolution. Its unique combination of natural resources also led it to produce Britain's first iron rails, iron bridge, iron boat and steam locomotive.

4.1 Just the job

For some jobs, you need just the right material, with the right combination of properties. Here we look at some interesting objects and the materials, both natural and synthetic, from which they are made.

Superconductors

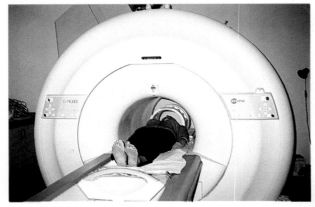

In this body scanner, the patient lies surrounded by a powerful electromagnet. The magnet is a coil of superconducting wire. When cooled to within four degrees of absolute zero, the wire loses all its electrical resistance. Then the current can flow without wasting any electrical energy, and a strong magnetic field is produced.

Fibre-reinforced polymers

The chassis of this racing car is made from plastic strengthened with carbon fibres. This makes it lighter and stronger than a metal chassis. Fibre-reinforced polymers can be used to make complex structures. A new design can take less than two weeks to get from the designer's screen to the Grand Prix race track.

Optical fibres

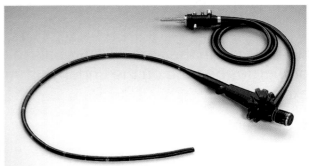

A flexible endoscope uses optical fibres, fine threads of pure silica glass, to allow the surgeon to see inside the patient. One set of fibres carries light into the body, while others transmit an image back to the surgeon.

Diamond

A carefully cut diamond enchants the eye. Diamond is simply pure carbon—the same atoms as in soot or graphite. Without impurities, diamond is clear and colourless, and because it is highly refractive and dispersive, white light reflects around inside and emerges with all the colours of the spectrum.

Yew wood

This ancient spear was found at Clacton in Essex. At 250 000 years old, it is one of the oldest wooden artefacts ever found. Yew wood is extremely hard and, when sharpened to a point, it would have been capable of penetrating the thick hide of a bison or even a rhinoceros. Its hardness also helped to preserve it down the long millennia.

Silicon solar cells

A solar battery can provide a clean, free-to-run source of electricity in a remote area. It can power a pharmacy's fridge for storing medicines, or a communal television set. Silicon is a semiconductor; some of the electrons in silicon are only weakly bound to atoms, and the energy of light is enough to set them free to carry an electric current.

One old bone

The human tibia found at Boxgrove.

Scientists and engineers aren't the only people who need to know about materials. This is the story of some ancient materials and the archaeologists who struggled to understand them.

In 1993, archaeologists working at Boxgrove in Sussex found part of a human bone, estimated to be half a million years old. It was half of a tibia, one of the bones in the lower leg. Later, they found a single tooth. Alongside these finds, they uncovered many tools and the remains of butchered animals. But what can you say about a group of people when all you have is one tooth and half a leg bone?

The femur, the single bone in the thigh, is the thickest bone in your body. It has to bear the weight of your body pressing down from above. And when you run, there is a large upward force on the femur

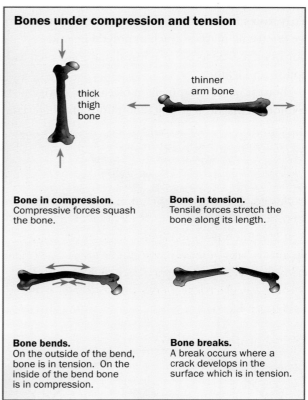

Bones under compression and tension

thick thigh bone

thinner arm bone

Bone in compression. Compressive forces squash the bone.

Bone in tension. Tensile forces stretch the bone along its length.

Bone bends. On the outside of the bend, bone is in tension. On the inside of the bend bone is in compression.

Bone breaks. A break occurs where a crack develops in the surface which is in tension.

every time your foot hits the ground, perhaps three times your bodyweight if you are sprinting. So the femur must be strong in compression. If you do a lot of running around, your body responds by building even thicker, stronger bones.

Detailed measurements of the Boxgrove tibia showed that it was remarkably thick and strong, much more so than the tibia of a modern human. This suggested that it came from a tall, powerful individual who led a very active hunting life.

The stuff of skeletons

Bone, the material from which all our bones are made, is a remarkable substance. In fact, it is a **composite material**, made from two very different substances:

- crystals of calcium phosphate, a brittle ceramic material, strong in compression, rather like chalk;
- fibres of collagen, a polymer made of protein, strong in tension, rather like rubber.

These materials belong to two different classes: ceramics and polymers. **Ceramic materials** are hard and brittle; they include china and other pottery, as well as more modern 'engineering ceramics', such as alumina and silicon carbide. **Polymers** include the familiar synthetic materials we often call 'plastics', such as polythene and polyester, as well as many natural materials such as cotton and leather.

Bone is nature's equivalent of carbon-fibre reinforced plastic, a strong but lightweight material.

Composite materials are designed to combine the most desirable properties of two or more different materials.

Tough tools

The Boxgrove archaeologists found that the people who lived there half a million years ago used a variety of tools. They knapped flint to make sharp knives and hand-axes. This is very skilled work. You need to know just where to strike a lump of flint to remove flakes so as to leave a useful core. Flint is a mineral, largely silica (silicon dioxide, SiO_2), similar to glass. One wrong blow and the hand-axe shatters and is wasted. The people of Boxgrove knapped the flint using harder stones; they also used softer hammers made of antler bone. They probably had

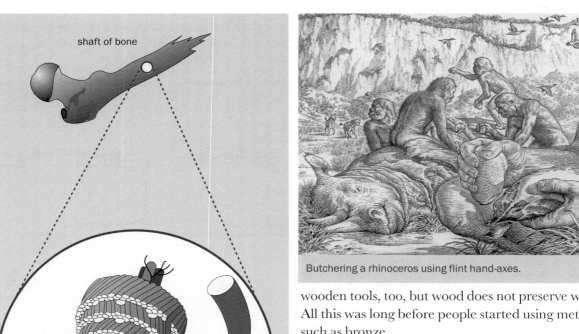

Butchering a rhinoceros using flint hand-axes.

wooden tools, too, but wood does not preserve well. All this was long before people started using metals such as bronze.

The purpose of these 'hand-axes' has been hotly debated. At Boxgrove, a local butcher was asked to cut up a carcass using a newly made axe. He found that the edge was sharp enough to give the clean cuts needed to do the job. When ancient axe-heads were examined under the microscope, they were found to have wear markings just like those on the one used by the butcher.

Sharp edges are characteristic of brittle, ceramic materials like flint. Ceramic materials, including china, shatter under impact rather than deforming. At the same time, these materials can be very hard, so that they do not squash or dent. A bone hammer,

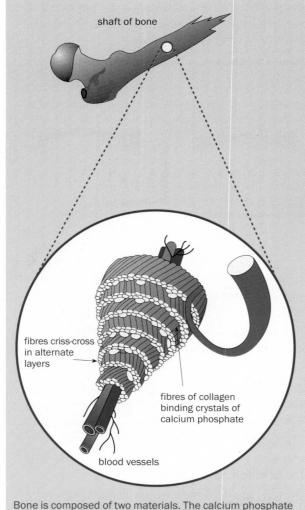

shaft of bone

fibres criss-cross in alternate layers

fibres of collagen binding crystals of calcium phosphate

blood vessels

Bone is composed of two materials. The calcium phosphate crystals are embedded in a mass of protective, slightly flexible collagen fibres. For added strength, the fibres in alternate layers are packed in a criss-cross fashion.

on the other hand, is softer. It squashes slightly when struck. Its flexibility means that, if you hit a bone hammer with great force, it is more likely to bounce than break.

One of the lessons of the Boxgrove archaeological dig, and many others at stone age sites, is that people have long used the materials they found around them to fashion useful artefacts. Today's world is very different. When the Boxgrove people killed a rhinoceros, they looked around for suitable flints to make hand-axes for their butchery work. Half a million years later, we have ready access to artefacts made from a vast range of materials which come to us from all over the world.

Brittle bones

As we age, our bones undergo an ageing process too. They tend to become more porous, and the proportion of calcium phosphate increases. Both of these factors make them more brittle. That's why elderly people are particularly vulnerable to falls, in which they can break a hip, leg or arm bone.

Brittle materials like glass or ageing bone fracture rather readily. When tests were carried out on the femurs of people who had died between the ages of three and ninety, it was found that the youngest bones absorbed three times as much energy as the oldest, before they broke. Brittleness is actually caused by cracks spreading through the material. The opposite of brittleness is **toughness**. In young bones the collagen fibres help to stop further cracks developing, and the young bones are tough. In old bones, cracks travel rather easily through the brittle calcium phosphate, present in larger amount. Engineers measure how easily a material fractures by measuring the energy absorbed when a prepared specimen is broken. To measure toughness you need to know how hard it is for cracks to deepen and spread.

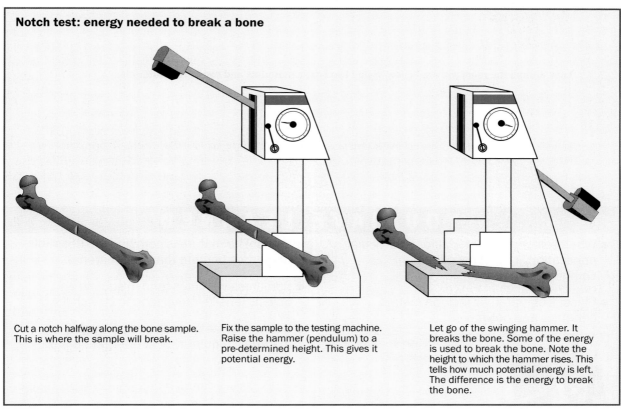

Notch test: energy needed to break a bone

Cut a notch halfway along the bone sample. This is where the sample will break.

Fix the sample to the testing machine. Raise the hammer (pendulum) to a pre-determined height. This gives it potential energy.

Let go of the swinging hammer. It breaks the bone. Some of the energy is used to break the bone. Note the height to which the hammer rises. This tells how much potential energy is left. The difference is the energy to break the bone.

● Materials from nature

Try these

1 Classify each of the following materials as metal, ceramic or polymer: flint, collagen, calcium phosphate, bronze, glass, rubber.

2 Give another example of a material from each class.

3 Give an example of a natural composite material and a synthetic composite.

4 Sketch diagrams to show the following:
● the bricks of a house are in compression
● a taut fishing line is in tension

5 Would you expect a shelf loaded with books to be in tension or compression?

6 Look around the room you are in now. Name two brittle materials and two tough materials.

Answers **1.** ceramic, polymer, ceramic, metal, ceramic, polymer **2.** for example: iron, china, polythene **3.** bone, carbon-fibre reinforced plastic **5.** upper surface in compression, lower surface in tension **6.** for example: glass, Perspex; steel, fibreglass

YOU HAVE LEARNED

● When selecting a material for a particular application, it must have the right combination of properties.

● Classes of materials include metals, ceramics and polymers.

ceramics, polymers, metals, composites, brittle, tough, compression, tension

● Composite materials combine the desirable properties of more than one material.

● Compressive forces tend to squash an object. Tensile forces tend to stretch it.

● Brittle materials break easily. Tough materials require a lot of energy to break them.

4.2 Better buildings

Tree house

Timber is a traditional building material in most parts of the world. It is excellent stuff; it has the stiffness and strength needed to make a building two or more storeys high. This is not surprising; timber comes from trees which themselves must be stiff and strong. A tall tree may be 30 m high and have a mass of 50 tonnes. It must be able to support its own weight, so the wood from which it is made must be strong in compression. At the same time, it must withstand high winds which try to bend it over, so it must also be strong in tension.

Wood is a natural composite material. It is composed of fibres of cellulose (the material of which the walls of plant cells are made), bound together by a substance called lignin. (Compare this with another natural composite, bone, described on pages 79–80.) The cellulose fibres are stiff and strong, and the lignin prevents them from splitting apart.

A traditional house in Malaysia is made from timber. Suitable timber is readily available from local forests, and techniques have been developed over centuries for cutting and joining sections of timber to make houses with several rooms. It's relatively easy to add an extension, and so a house can eventually become a maze of interconnecting rooms.

Tall storeys

In earlier decades, the height of skyscrapers was limited by their weight: the lower storeys could not withstand the stress of the upper storeys pressing down on them. There were problems too with swaying in high winds. For a long time, the Empire State Building in New York represented the limit of high-rise building. Now, developments both in materials (which are stiffer, stronger and more lightweight) and in design have resulted in a new generation of tall buildings.

The Malaysian capital Kuala Lumpur claims the world's tallest office building, the twin towers of the Petronas building. This 88 storey building, 452 m high and made of steel and glass, was completed in 1997. Built in Kuala Lumpur's city centre, it represents Malaysia's claim to be a major centre for world commerce.

- Creep in an everyday material
- Time effects in an everyday material

Choosing building materials

Modern buildings may be made from a great range of materials—stone, brick, steel, timber, glass, concrete. How can an architect set about deciding whether a particular material is suitable for the job? And what dimensions must a particular section have if it is to withstand the forces which will act on it?

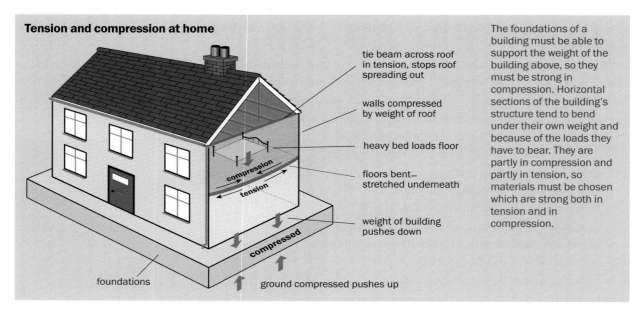

Tension and compression at home

tie beam across roof in tension, stops roof spreading out

walls compressed by weight of roof

heavy bed loads floor

floors bent—stretched underneath

weight of building pushes down

compression
tension
compressed

foundations

ground compressed pushes up

The foundations of a building must be able to support the weight of the building above, so they must be strong in compression. Horizontal sections of the building's structure tend to bend under their own weight and because of the loads they have to bear. They are partly in compression and partly in tension, so materials must be chosen which are strong both in tension and in compression.

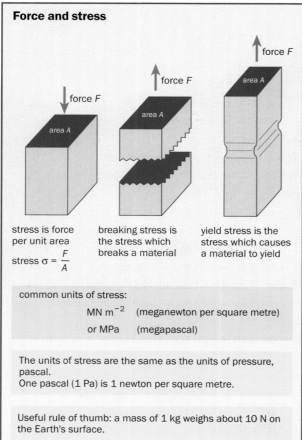

Force and stress

force F

force F

force F

area A

area A

area A

stress is force per unit area

stress $\sigma = \dfrac{F}{A}$

breaking stress is the stress which breaks a material

yield stress is the stress which causes a material to yield

common units of stress:

    MN m^{-2} (meganewton per square metre)

    or MPa   (megapascal)

The units of stress are the same as the units of pressure, pascal.
One pascal (1 Pa) is 1 newton per square metre.

Useful rule of thumb: a mass of 1 kg weighs about 10 N on the Earth's surface.

Concrete is often used for the foundations of a building, and its strength in compression might be about 50 N mm^{-2}. This means that each square millimetre of concrete can withstand a force up to 50 N pressing down on it—that's the same as the weight of 5000 tonnes pressing down on each square metre. A greater force will crush the concrete and the building will be in danger of collapsing.

Timber is good for horizontal sections (think of the timber joists which are often used to support floors), and so are metals such as steel. Concrete and brick are less good because they are much weaker in tension than in compression.

Metals have a different problem—they 'give' before they break. As the load on a piece of steel is increased, it reaches a point at which the steel yields and becomes permanently deformed. This is the **elastic limit** or yield point, and the architect must ensure that the load on the steel is always less than this or it will bend and buckle. So it isn't the breaking stress (the amount of stress that breaks the metal) which determines its useful strength; what matters is the stress at the yield point. **Breaking stress** and **yield stress** are two ways of thinking about how strong a material is.

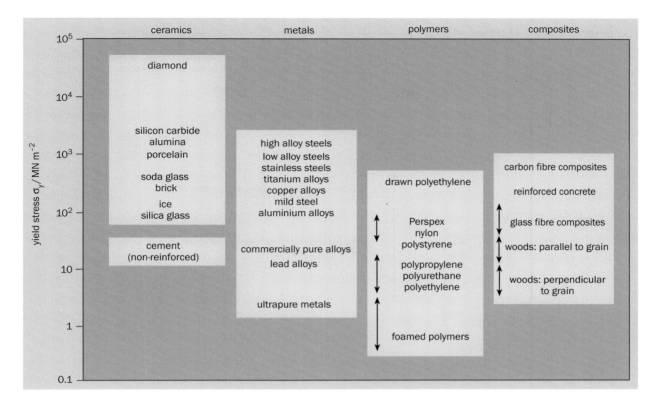

Spiders' webs and steel

It's often said that the threads of a spider's web are stronger than steel—and it's true. This seems surprising, because we know how easy it is to brush aside a spider's web, while steel is used because it is strong. But if we compare a spider's thread with a steel thread of the same thickness, we find that the steel is weaker.

This machine is being used to test a block of stone. The machine is controlled and monitored by a computer. Machines like this are used to measure the strengths of materials. Metals such as steel are tested by stretching them in a tensile testing machine.

By comparing threads of the same thickness (or cross-sectional area), we are ensuring a fair test of the two materials. It's true that a spider's web is weaker than a steel bar, but the material that it is made from is stronger than steel. We must distinguish between the strength of an object and the strength of the material from which it is made.

The chart above shows the strengths of a number of different materials, grouped according to their classes. Things to note about this chart:

- The left-hand scale is logarithmic. Each step is a factor of ten.
- Strength is represented as yield stress, σ_y in $MN\,m^{-2}$.
- The strongest material (diamond) is at least 10^5 times as strong as the weakest (foamed polymers).
- Cement is a weak material, but when it is made into concrete and reinforced by adding steel rods, it becomes much stronger.

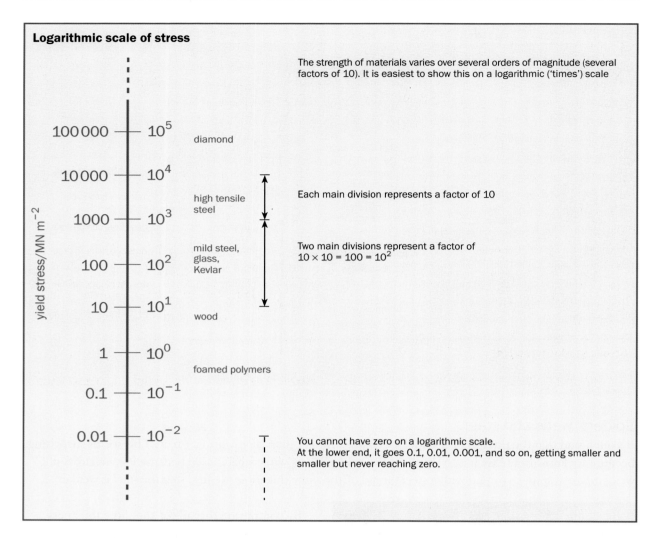

Logarithmic scale of stress

The strength of materials varies over several orders of magnitude (several factors of 10). It is easiest to show this on a logarithmic ('times') scale

yield stress/MN m^{-2}

100 000	10^5	diamond
10 000	10^4	high tensile steel
1000	10^3	
100	10^2	mild steel, glass, Kevlar
10	10^1	wood
1	10^0	foamed polymers
0.1	10^{-1}	
0.01	10^{-2}	

Each main division represents a factor of 10

Two main divisions represent a factor of $10 \times 10 = 100 = 10^2$

You cannot have zero on a logarithmic scale. At the lower end, it goes 0.1, 0.01, 0.001, and so on, getting smaller and smaller but never reaching zero.

Surviving an earthquake

During an earthquake, a building may shake back and forth so vigorously that it falls apart. The forces in it exceed the breaking strength of the material from which it is constructed. We don't generally want to live in buildings that bend back and forth, but some flexibility is inevitable. In an earthquake zone, it may even help by absorbing the energy of a 'quake and leave the building standing.

Building materials must be stiff—no-one wants to walk around on floors that stretch like trampolines. We don't want window panes that bow inwards when the wind blows, or outwards when we sneeze. But all materials have some degree of flexibility. The same testing machine which measures the strength of a material can be used to find out how flexible or stretchy it is; the result is known as the **Young modulus** of the material, though engineers often call it simply the elastic modulus. Knowing the Young modulus, designers can calculate how thick a beam or other building component must be to support the weight of any load it is likely to have to bear, without bending too much.

Rubber and jelly are examples of materials at the opposite end of the scale of stiffness. They are flexible, and bend or stretch very easily. Plastics like polythene are a bit stiffer.

● More about log scales

This Japanese building was designed to be strong. It has remained in one piece despite having been toppled by an earthquake.

Stress produces strain

How much does a material stretch? For a given force F a long piece of material will stretch more than a short one. Doubling the length L doubles the extension x

strain $\dfrac{x}{L}$

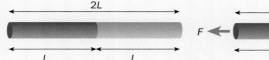

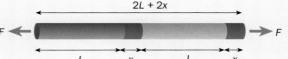

strain $\dfrac{2x}{2L} = \dfrac{x}{L}$

extension x depends on original length
strain = fractional increase in length and does not depend on the original length

$$\text{strain } \varepsilon = \frac{x}{L}$$

Units: strain compares lengths. If the units of extension and length are the same, strain ε is simply a number. It is often reported as a percentage, e.g. 1%

● Hooke's law and the Young modulus

● Materials database

Breaking glass, stretching steel

It can be hard to believe that glass is a flexible material. If you press gently on the middle of a window pane, it will bow outwards slightly. And think of the glass fibres of a surgeon's endoscope—they can bend, following the curves of the patient's insides.

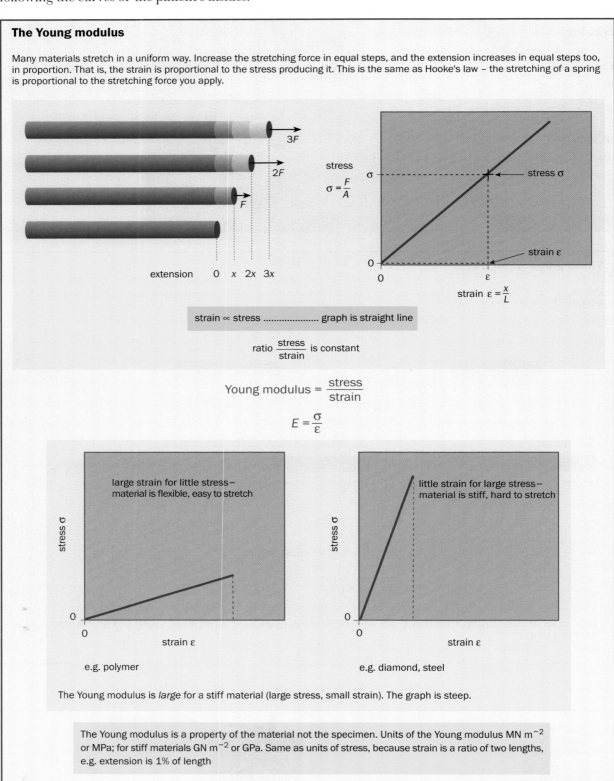

The Young modulus

Many materials stretch in a uniform way. Increase the stretching force in equal steps, and the extension increases in equal steps too, in proportion. That is, the strain is proportional to the stress producing it. This is the same as Hooke's law – the stretching of a spring is proportional to the stretching force you apply.

$$\sigma = \frac{F}{A}$$

$$\text{strain } \varepsilon = \frac{x}{L}$$

strain ∝ stress .... graph is straight line

ratio $\dfrac{\text{stress}}{\text{strain}}$ is constant

$$\text{Young modulus} = \frac{\text{stress}}{\text{strain}}$$

$$E = \frac{\sigma}{\varepsilon}$$

large strain for little stress— material is flexible, easy to stretch

little strain for large stress— material is stiff, hard to stretch

e.g. polymer

e.g. diamond, steel

The Young modulus is *large* for a stiff material (large stress, small strain). The graph is steep.

The Young modulus is a property of the material not the specimen. Units of the Young modulus MN m^{-2} or MPa; for stiff materials GN m^{-2} or GPa. Same as units of stress, because strain is a ratio of two lengths, e.g. extension is 1% of length

A small stress will cause glass to stretch slightly. Remove the stress and it returns to its original dimensions. This is **elastic deformation**. Increase the stress too far and you pass the point of no return—the glass breaks. This is **brittle fracture**.

Some metals are brittle—they behave like glass. Others, such as mild steel, are different. Think of what happens when a car crashes. The steel bodywork is easily dented, but it is unlikely to break.

For very small strains, mild steel shows elastic behaviour, like a spring. But then, at the elastic limit, it yields. Now it begins to deform—this is **plastic deformation**. The strain may increase to 40% or more before it eventually fractures.

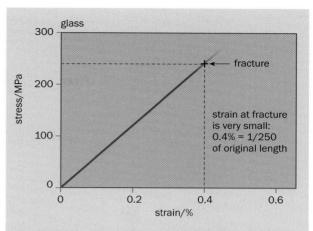

The stress–strain graph for glass is a straight line, up to the point where it fractures. Note that the strain when it breaks is less than 1%—its length increases by this tiny amount.

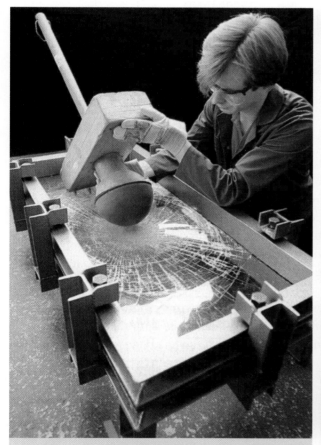

Beyond the elastic limit—testing window glass at the British Standards Institute. The wooden testing device represents a human head.

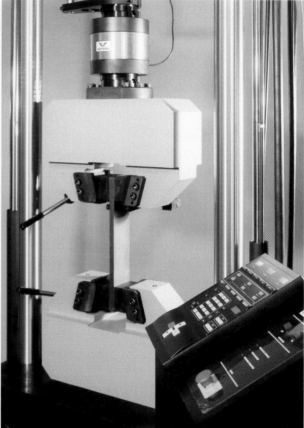

Specially shaped metal samples are put in this tensile testing machine. They are gradually stretched—the strain increases at a steady rate.

● Fantastic fibres

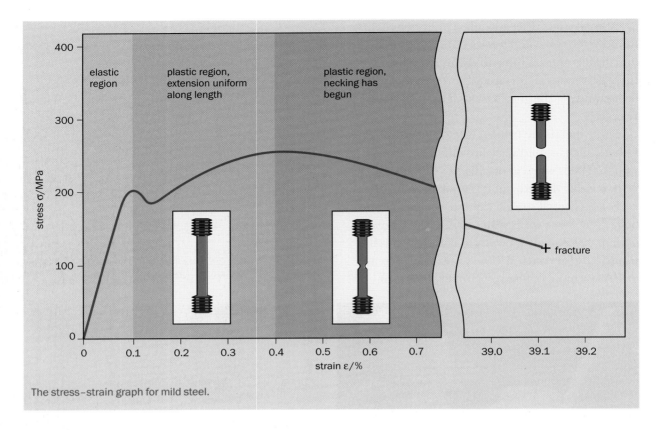

The stress–strain graph for mild steel.

Wood ready to wear

This section started with wood for buildings, and ends with wood as a source of fibres. A family of four can be provided with a lifetime of clothing by just one tree. But for this to work, you would really have to like viscose—the material which is manufactured from the bark of eucalyptus, pine or beech trees. Viscose, and its close relations acetate and triacetate, are all thin, soft, slightly shiny fabrics which are often used to make dresses and to line suits. But just imagine having all your clothes—underwear, socks, trousers, jackets and skirts—made from just this one material.

Viscose is a polymer. The fact that viscose clothing will fit around your body shows that it has a degree of elasticity. In fact, polymers tend to be more elastic than most metals or ceramics. But don't be misled by the way we often refer to some polymers as 'plastics'. Although some polymers show plastic behaviour, others are brittle. And their behaviour can change with temperature. Perspex is brittle, like glass, at room temperature. Warm it up and it becomes bendy; warm it some more and you can deform it to any shape you like.

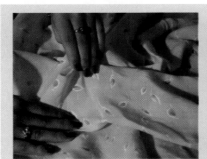

Almost all clothing is made from polymers. We rely on their flexibility for a figure-hugging fit. Others must be stiff for things like shoe soles. Even the natural materials we wear—cotton, leather—are polymers.

Try these

The first four questions relate to the chart on page 85.

1   **Which class of materials is generally the strongest?**

2   **Which are stronger, pure metals or alloys?**

3   **'Ice' is a slang name for diamonds. How much stronger is diamond than ice?**

4   **How much stronger is reinforced concrete than cement which has not been reinforced?**

5   **A concrete block of cross-sectional area 0.01 m^2 is crushed when a force of 1 MN is applied to it. What is the breaking stress in MN m^{-2}? And in N mm^{-2}? How many kilograms can it support? (Recall that 1 kg weighs about 10 N.)**

6   **Look at the stress–strain graph for glass on page 89. From the graph, estimate the breaking stress and the Young modulus of glass. Do the same for mild steel, using the graph on page 90.**

Answers 1. ceramics **2.** alloys **3.** 10^3 times **4.** 10 times **5.** 100 MN m^{-2}, 100 N mm^{-2}, 100 000 kg
6. 240 MPa, 60 GPa, 125 MPa, 200 GPa

YOU HAVE LEARNED

- Ceramic materials such as brick and concrete are strong in compression but weak in tension.

- The strength of a material is represented by its breaking stress or yield stress.

- The stiffness of a material is represented by its Young modulus.

- The Young modulus is found from the initial gradient of a stress–strain graph.

- Materials are elastic up to the elastic limit; then they either fracture or show plastic deformation.

- $\text{stress} = \dfrac{\text{force}}{\text{area}}$

$\text{strain} = \dfrac{\text{extension}}{\text{original length}}$

$\text{Young modulus} = \dfrac{\text{stress}}{\text{strain}}$

stress, strain, Young modulus, yield stress, tensile strength, brittle, tough, elastic

4.3 Crystal clear?

Colourless glass can make a colourful sculpture. Where does the colour come from? Reflection and refraction of light play a part. So do transmission, absorption and dispersion. It's a complicated business, and artists working with glass have learned from experience how to achieve the effects they desire. Physicists try to explain these effects.

It all comes down to a question of the speed of light. Light travels fastest in a vacuum, at almost $3 \times 10^8 \text{ m s}^{-1}$. It travels only slightly slower in air (because air is mostly a vacuum). But when light travels from air into glass, it slows down a lot. This is because light interacts with the electrons in the

Glass sculpture 'Space' (1980) by the Czech artist Pavel Hlava. Light is reflected and refracted in the interior of the sculpture.

atoms of the glass. From the refractive index n of the glass we can work out how much the light slows down. This depends on the composition of the glass—there isn't just one type of glass. Artists have traditionally used lead crystal glass for their finest work; this has a high refractive index, and so has the biggest effect on the light.

There are two ways in which white light can become coloured light. Recall that white light is a mixture of all the colours of the spectrum, from red to violet. When white light passes through coloured glass, some of these colours are absorbed. So when white light passes through glass which absorbs red light, it comes out looking blue-green.

The second way depends on refraction. When light enters glass, some colours slow down more than others—the short wavelength violet light is affected more than the longer wavelength red. So the violet light bends more than the red, and we end up with a spectrum—all the colours of the rainbow.

We think of 'white light' as being uncoloured. It might be better to think of 'white' as the colour of light we see direct from the Sun. We see other colours when some wavelengths of that light are removed.

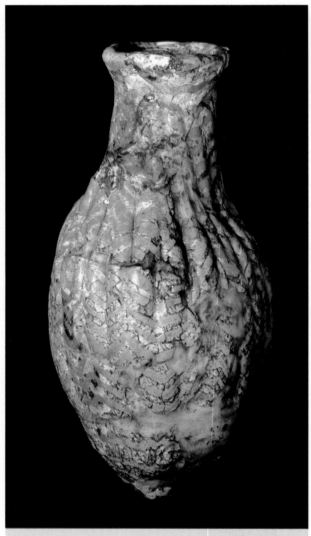

Glass bottle, made about 1300 BC, found in Ur (in present-day Iraq) near where glass-making seems to have begun about 3000 BC.

Defining refractive index

Light slows down when it enters glass, and speeds up again as it leaves. The wavefronts are closer together in glass than in air. (You may have already used this way of representing light in chapter 1.)

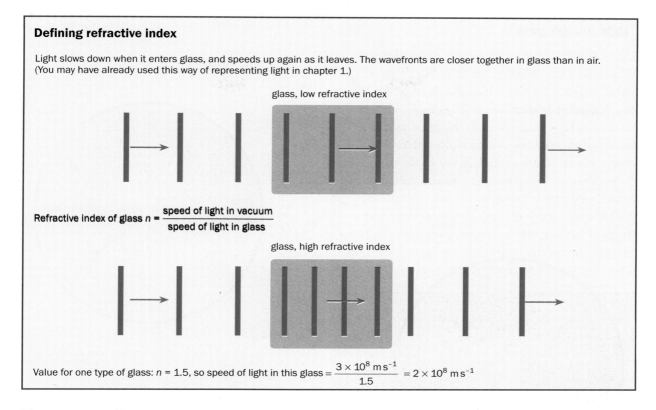

glass, low refractive index

Refractive index of glass $n = \dfrac{\text{speed of light in vacuum}}{\text{speed of light in glass}}$

glass, high refractive index

Value for one type of glass: $n = 1.5$, so speed of light in this glass $= \dfrac{3 \times 10^8 \, \text{m s}^{-1}}{1.5} = 2 \times 10^8 \, \text{m s}^{-1}$

Now you see it ...

Here's one way to find the refractive index of a piece of glass. Immerse it in some clear oil. If oil and glass have the same refractive index, the glass will be virtually invisible.

How it works: when the two refractive indices match, the light doesn't change speed as it passes from oil to glass. There is no reflection or refraction, so we can't see the glass.

Forensic scientists use this technique to identify fragments of glass found at the scene of a crime.

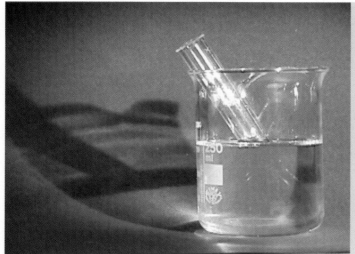

This photo shows a Pyrex boiling tube immersed in glycerol. At the right temperature, the refractive indexes of the two materials are identical. Not only does the tube look invisible in the beaker, it casts no shadow.

Mineralogists keep sets of calibrated oils in their laboratories, which cover a range of refractive indices. They can identify transparent crushed samples of minerals under the microscope, finding which oil makes the fragments vanish when a drop of the oil is put on them. They can even mix the oils to obtain greater precision, taking an average of the refractive indices of the oils in the mixture.

● More about Snell's law

Light through glass

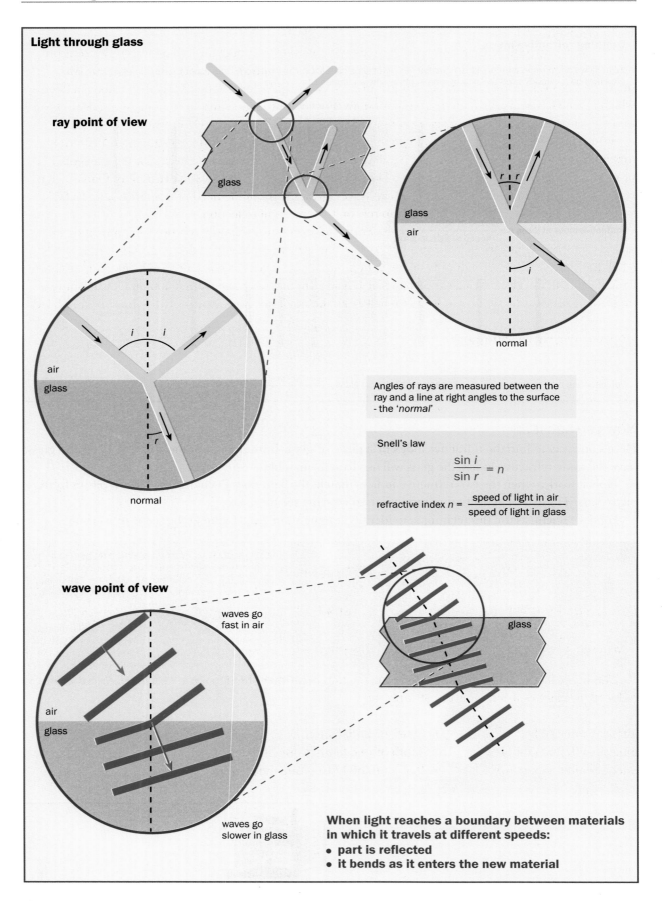

ray point of view

glass

air

glass

normal

r r

glass

air

i

normal

i i

r

Angles of rays are measured between the ray and a line at right angles to the surface - the 'normal'

Snell's law

$$\frac{\sin i}{\sin r} = n$$

refractive index $n = \dfrac{\text{speed of light in air}}{\text{speed of light in glass}}$

wave point of view

waves go fast in air

air

glass

glass

waves go slower in glass

When light reaches a boundary between materials in which it travels at different speeds:
- **part is reflected**
- **it bends as it enters the new material**

Pure thoughts

Doctors and surgeons now routinely look inside their patients, to inspect the patients' digestive or reproductive systems. They use an instrument called an endoscope—one is shown in the photograph on page 78. The endoscope consists of a bundle of high-purity glass fibres, bound together to form a device which can be safely inserted into the patient's body. Light can be shone into the patient and reflected back to the doctor's eye. Rays of light zig-zag up and down the fibres by means of total internal reflection.

When you make a long-distance telephone call, or receive cable television, you are almost certainly making use of optical fibres. Fibre optic systems have rapidly replaced copper wires for the transmission of electronic signals for two main reasons: more information can be sent down a glass fibre than down a wire of the same thickness, and the signal can travel further without the need of amplification. Today's fibres are so thin that they no longer need to rely on total internal reflection.

High-purity glass is at the heart of the fibre optic telecommunications revolution. An optical signal (in effect, a flashing infrared light) is passed down a fine glass fibre. But even the best quality glass won't let all the light through; the intensity of the light gradually decreases as it is absorbed by any impurities or irregularities in the glass. Using silica glass which is so pure that only one atom in a billion of impurity is present means that the signal can travel up to 100 km before it becomes so weak that it must be amplified.

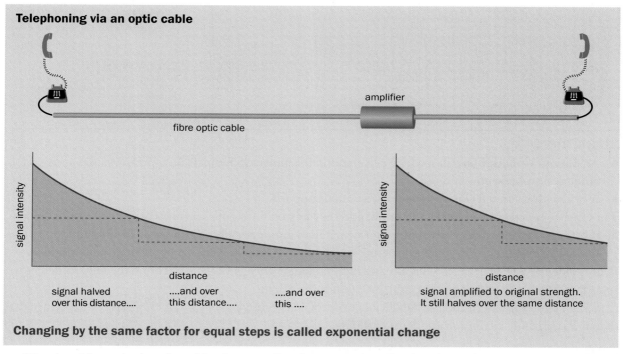

Telephoning via an optic cable

amplifier

fibre optic cable

signal intensity

distance

signal halved over this distance....

....and over this distance....

....and over this ....

signal intensity

distance

signal amplified to original strength. It still halves over the same distance

Changing by the same factor for equal steps is called exponential change

The signal intensity is reduced by the same fraction every time the signal travels a given distance. If it is reduced by a factor of ten for every 30 km:

intensity after 30 km: 10%	one tenth of initial intensity
intensity after 60 km: 1%	one tenth of one tenth = one hundredth
intensity after 90 km: 0.1%	1/10 of 1/10 of 1/10 = one thousandth

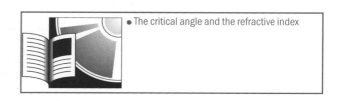

● The critical angle and the refractive index

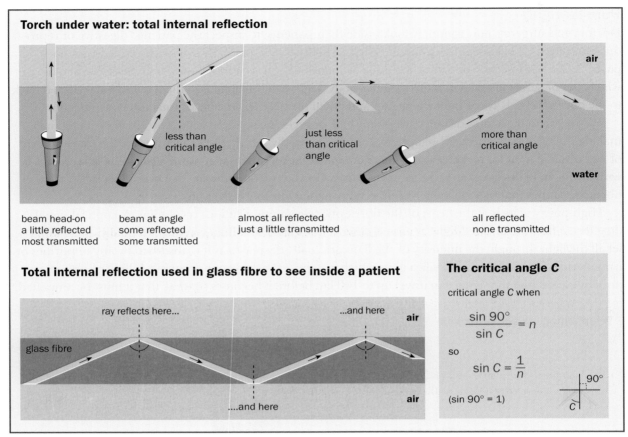

Torch under water: total internal reflection

air

less than
critical angle

just less
than critical
angle

more than
critical angle

water

beam head-on	beam at angle	almost all reflected	all reflected
a little reflected	some reflected	just a little transmitted	none transmitted
most transmitted	some transmitted		

Total internal reflection used in glass fibre to see inside a patient

ray reflects here...   ...and here
air

glass fibre

....and here
air

The critical angle C

critical angle C when

$$\frac{\sin 90°}{\sin C} = n$$

so

$$\sin C = \frac{1}{n}$$

(sin 90° = 1)   90°

C

Smart materials

- An aircraft with a surface skin which can respond to changes in temperature, pressure, air speed and other atmospheric conditions.
- A bridge that can tell you the stresses in its main supports.

These are two examples of smart materials, and they rely on fibre optics. Fibres carry signals from sensors on the aircraft's wings to a central computer which then automatically adjusts the instrumentation.

Optical fibres can detect when a structure is damaged. A loop of fibre carries light around a bridge. If the bridge is cracked, the fibre breaks and light no longer travels all the way round the loop. A detector reports the decrease in transmitted light and sounds the alarm.

Optical fibres are becoming the artificial nerves of our technology. One day they may even replace the damaged nerves of victims of accidents or illness.

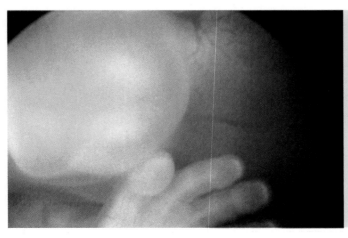

Images like this, captured using an optical-fibre endoscope, have helped us to understand much more about the development of a new life in the womb. An endoscope is essentially a bundle of optical fibres surrounded by a protective and flexible sheath. Some fibres carry light into the body and others return light to the operator. Endoscopy is now routinely used for medical diagnoses, for example looking at organs in the digestive tract. It is also used in 'keyhole' surgery.

Try these

1 Pyrex glass has a refractive index *n* of 1.47. For window glass, *n* = 1.51. In which does light travel faster?

2 The refractive index of water is 1.33, or 4/3. What is the speed of light in water? (Speed of light in vacuum $c = 3 \times 10^8$ m s^{-1}.)

3 Sketch a diagram to show why rays of light are scarcely affected as they pass through a thin pane of window glass.

4 When does a refracted ray of light bend *away from* the normal?

5 A light signal is sent along a 100 km long fibre optic cable. At the far end, its intensity is just 1% of its initial value. What would its intensity be at the end of a 200 km cable?

6 What would its intensity be halfway along the 100 km cable?

Answers **1.** Pyrex **2.** 2.25×10^8 m s^{-1} **4.** when entering a material with lower refractive index, e.g. from glass to air **5.** 0.01% **6.** 10%

YOU HAVE LEARNED

- When light enters a transparent material, some may be reflected and some refracted.

- Light slows down when entering a transparent material; the refractive index shows how much it slows:

  Refractive index of material
  $$n = \frac{\text{speed of light in vacuum}}{\text{speed of light in material}}$$

- The refractive index of the material determines the angle through which the light is refracted; it also determines the fraction which is reflected.

- Snell's law: sin *i* / sin *r* = *n*

- Very high purity glass is used for optical fibres.

- Light may be transmitted through an optical fibre by total internal reflection.

- Optical fibres can carry a higher density of information than copper cables, and with less frequent need for amplification.

refractive index, exponential change, normal, critical angle, total internal reflection

4.4 Conducting very well—conducting very badly

Some materials are better electrical conductors than others. The scale of conductivity below shows that some are much, much better conductors. The best conductor (silver) has a conductivity which is almost a million million million million (10^{24}) times greater than that of the best insulator (polystyrene). Electrical conductivity varies over a wider range than any other material property. Because of the great range of values, the chart uses a logarithmic ('times') scale.

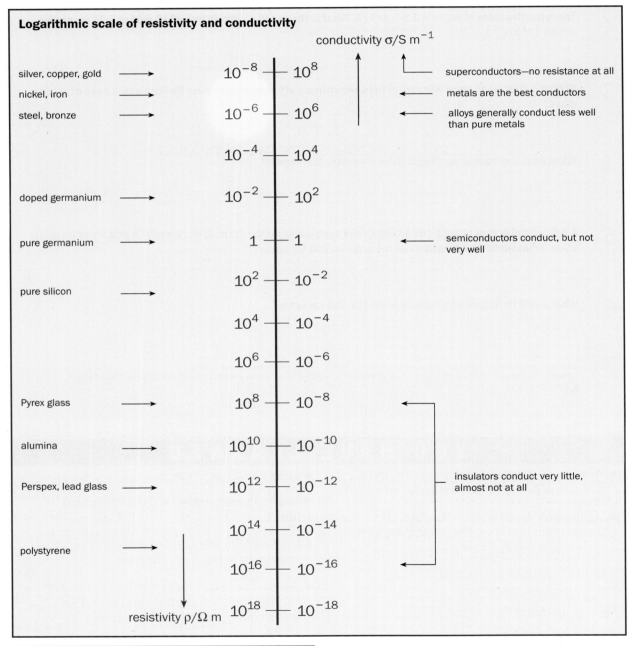

Logarithmic scale of resistivity and conductivity

conductivity σ/S m^{-1}

silver, copper, gold →	10^{-8} — 10^{8}	superconductors—no resistance at all
nickel, iron →		metals are the best conductors
steel, bronze →	10^{-6} — 10^{6}	alloys generally conduct less well than pure metals
	10^{-4} — 10^{4}	
doped germanium →	10^{-2} — 10^{2}	
pure germanium →	1 — 1	semiconductors conduct, but not very well
pure silicon →	10^{2} — 10^{-2}	
	10^{4} — 10^{-4}	
	10^{6} — 10^{-6}	
Pyrex glass →	10^{8} — 10^{-8}	
alumina →	10^{10} — 10^{-10}	
Perspex, lead glass →	10^{12} — 10^{-12}	insulators conduct very little, almost not at all
	10^{14} — 10^{-14}	
polystyrene →	10^{16} — 10^{-16}	
	10^{18} — 10^{-18}	

resistivity ρ/Ω m

● Resistivity and conductivity calculations

Conductivity and resistivity

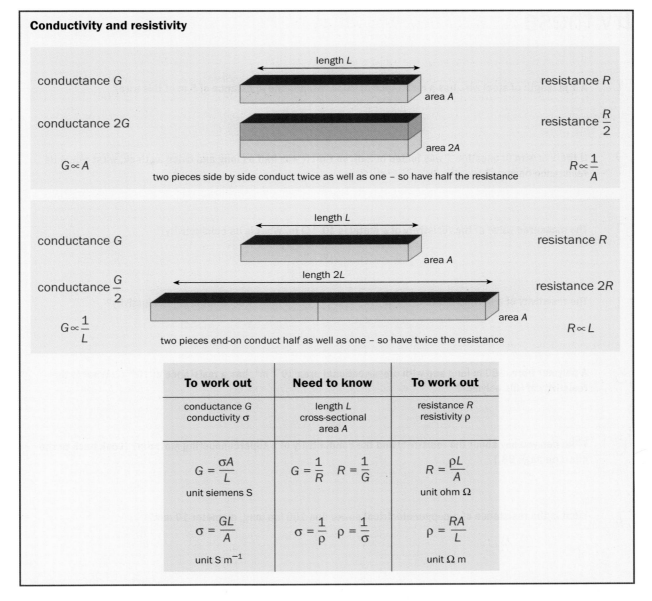

To work out	Need to know	To work out
conductance G conductivity σ	length L cross-sectional area A	resistance R resistivity ρ
$G = \dfrac{\sigma A}{L}$ unit siemens S	$G = \dfrac{1}{R}$ $\quad$ $R = \dfrac{1}{G}$	$R = \dfrac{\rho L}{A}$ unit ohm Ω
$\sigma = \dfrac{GL}{A}$ unit S m^{-1}	$\sigma = \dfrac{1}{\rho}$ $\quad$ $\rho = \dfrac{1}{\sigma}$	$\rho = \dfrac{RA}{L}$ unit Ω m

Materials and artefacts

Which is heavier, a ton of lead or a ton of feathers? Of course, both have the same mass but lead has the greater density. In this chapter, you have seen several examples of how to distinguish between the properties of an individual object or artefact, and the properties of the materials from which it is made. Mass and density are just one pair of quantities where care is needed. The mass of an object depends on its dimensions as well as the material it is made from.

property of artefact	Unit	property of material	Unit
mass (of an object)	kg	density	kg m^{-3}
conductance (of a wire)	S	conductivity	S m^{-1}
resistance (of a wire)	Ω	resistivity	Ω m
stiffness (of a spring)	N m^{-1}	Young modulus	N m^{-2}
force to permanently deform (a metal bar)	N	yield stress	N m^{-2}
force to break (a window)	N	breaking stress	N m^{-2}

Try these

1   A 1 m length of steel wire has a resistance of 20 Ω. What is the resistance of 5 m of this wire?

2   If the 1 m wire in question 1 was folded in half, so that it was half as long and twice as thick, what would its resistance become?

3   The measured value of the resistivity of a metal is 10^{-8} Ω m. What is its conductivity?

4   The resistivity of a second metal is 2×10^{-8} Ω m, twice that of the first. What is its conductivity?

5   A polymer fibre, 100 m long and with cross-sectional area 10^{-6} m², has a resistance of 10^{18} Ω. What is the resistivity of this polymer?

6   What can you say about the resistivity and the conductivity of a superconducting material? (Look back to the chart on page 98.)

7   What is the resistance of a copper electrical power line 100 km long, diameter 10 mm?

Answers 1. 100 Ω **2.** 5 Ω **3.** 10^8 S m^{-1} **4.** 0.5×10^8 S m^{-1} **5.** 10^{10} Ω m **6.** zero resistivity, infinite conductivity **7.** about 13 Ω

YOU HAVE LEARNED

- Electrical conductivity is a property which varies over a vast range.

resistivity, electrical conductivity

- It is important to distinguish between the properties of an artefact (e.g. resistance) and those of a material (e.g. resistivity).

- In selecting a material for a particular application, many different properties may have to be taken into account.

Summary checkup

✓ Classes of materials:

- Classes of materials include metals, ceramics and polymers.

- Composites are designed to combine desirable properties from two or more materials.

✓ Elastic behaviour:

- Solid materials behave elastically up to the elastic limit.

- The Young modulus is the gradient of the slope of the initial, linear part of the stress–strain graph.

- Young modulus = stress/strain

   where stress = load/area, strain = increase in length/initial length

✓ How materials fail:

- Metals deform plastically before they break. A lot of energy is needed to break them—they are tough materials.

- Brittle materials (notably ceramics, including glass) break easily, without deforming plastically.

- Some polymers are brittle; some are very elastic; some are very tough.

✓ Optical properties:

- When entering a medium of greater refractive index, a ray of light is partly reflected; the refracted ray bends towards the normal.

- Refractive index = speed of light in a vacuum/speed of light in material

- The direction of the refracted ray is given by Snell's law: $\sin i \, / \sin r = n$

✓ Electrical properties:

- Electrical conductivity tells you how well a material conducts an electric current.

- Electrical conductivity varies very widely between materials.

- Electrical resistivity ρ is the reciprocal of conductivity σ: $\rho = 1/\sigma$

- To calculate the resistance R of an object knowing the resistivity ρ of the material from which it is made, use $R = \rho L/A$

Questions

1 Most cups or mugs used at home are made from china; picnic cups are usually plastic.

   (a) Explain why these different materials are used; refer to these properties: density, stiffness, brittleness, and any other properties which you consider important.

   (b) What key properties would be important for a material used for dental fillings? Name two materials commonly used for this purpose.

2 Look at the five objects A–E. Each is being pulled by the force shown. Explain why B will be stretched more than A, C more than B, D more than C, and E more than D.

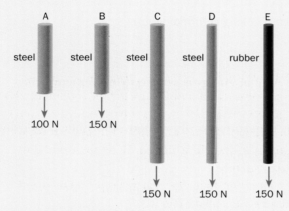

3 A rod is made from a type of glass whose breaking stress is 40 MPa. The cross-sectional area of the rod is 100 mm² (10^{-4} m²).

   (a) What is meant by the term *breaking stress*?

   (b) What force is needed to break the rod?

4 Steel has conductivity 10^6 S m^{-1}, Young modulus 200 GPa, yield stress 200 MPa.

   (a) Estimate the radius and cross-section of an unstretched 1 m long steel wire with resistance $\frac{1}{3}$ Ω.

   (b) Estimate the tension in this wire when it yields.

   (c) Assuming that the extension is linear up to the yield point, estimate by how much the wire has stretched just before it starts to yield.

   (d) Why does the resistance of the wire change as it is stretched? Will the change be large?

5 A rubber cord, 5 m long and 50 mm thick, hangs from the top of a zoo cage. A chimp weighing 300 N hangs on to the end of the cord. How long is the cord now? (The Young modulus of the rubber = 100 MPa.)

6 The diagram shows a ray of light entering a block of Perspex (refractive index $n = 1.5$).

   (a) What is the angle of incidence?

   (b) Calculate the angle of refraction.

   (c) What is the speed of light inside the block?

7 The refractive index for diamond is about 2.5, whereas the refractive indices for glass and water are similar (see page 93). Explain how you might use this information to check whether the diamond in a ring you are offered is genuine or fake.

8 The table shows the resistance of 1 m lengths of various wires, each 0.25 mm in diameter, and made from different materials.

material	resistance/Ω
copper	0.351
Eureka	10.0
manganin	8.45
Nichrome	22.0

   (a) Which of these materials has the greatest resistivity?

   (b) Calculate the resistivity of this material.

9 Use the chart of conductivity on page 98 to find how much better a conductor copper is than steel. How do *semiconductors* get their name?

5 Looking inside materials

Damascus steel for swords and Chinese porcelain for ornaments show how properties of materials have been improved by trial and error. Today, better understanding of how the structure of materials decides their properties leads to designer materials, with purpose-built structure providing desired properties. Old needs are met better, and new uses opened up. We will give examples of:

- how the structure of a material determines its properties

- how materials can be designed to have needed properties

- how different properties—mechanical, electrical and optical—can be controlled by the same structure

5.1 Materials under the microscope

How can you find out about the structures of materials? Look at a piece of wood. Its grain reveals its structure. Wood for furniture is usually cut so that the grain runs straight down the legs of a chair, or along the length of a table top. Take an ice lolly stick and split it end to end. The wood breaks easily between the fibres which make up the grain. Break the stick across its length and you will be left with rough ends, with stiff fibres exposed. Under an optical microscope, the long, narrow cells which make up the wood are revealed. It is these cells which give wood its characteristic structure.

So you can see the structure of wood. Carpenters pay attention to it, and will avoid pieces of wood with knots or distorted grain. These may introduce weakness into a chair leg.

But what about materials where the structure is not obvious? The metal of a coin or the plastic of a Lego brick may appear entirely uniform to the naked eye. Today there are several different types of microscope which can help to show the internal structures of such materials.

As physicists dissected the anatomy of atoms … all kinds of engineers and scientists delved in ever more detail into the relationship between a material's many-leveled structure and its properties. Why exactly do ceramic materials break so readily but metals only dent? Why does steel resist rusting when you put enough chromium into it? What is going on in a piece of metal when it undergoes age-hardening? What makes rubber so elastic? Can a carefully constructed crystalline material function like an entire device, say, like the vacuum tubes of early electronic technology? … Like a person's personality, every material's collection of traits and behaviours comes from what it is made of and how those ingredients were nurtured into their present form.

Ivan Amato, in *Stuff: The Materials the World is Made of* **(1997) Avon Books**

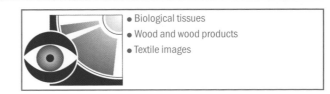

- Biological tissues
- Wood and wood products
- Textile images

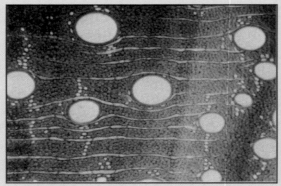

Viewed under an optical (light) microscope, wood is shown to be made of many long, parallel cells which give the material its characteristic grain structure.

An unusual image—an ant carrying a tiny electronic microchip, viewed with a scanning electron microscope (SEM). The advantage of the SEM over an optical microscope is that the whole insect can be in focus. The disadvantage is that the image is only in black and white.

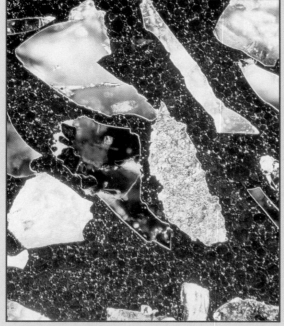

An optical microscope was used to make this colourful image of a 'cermet'—a composite material. Angular ceramic particles are embedded in steel.

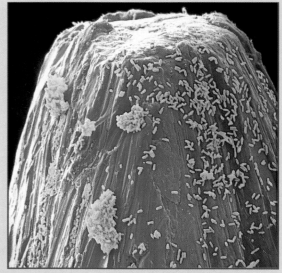

This is a false-colour SEM image of bacteria (yellow) on the point of a pin. It has been coloured to show the different features. Notice how the steel of the pin, which would feel smooth to the touch, is in fact rough when seen on this scale.

The scanning tunnelling microscope (STM—see page 11) allows us to see matter at the scale of individual atoms. Here, graphite atoms (green) form a regular array, with atoms of gold (orange) piled up on the graphite surface. Another false-colour image.

The atoms of a metal usually form a regular, crystalline array, as shown in this image, which was made using an atomic force microscope (AFM). The microscope can only just resolve the atoms. The conical appearance is an artefact of the imaging process.

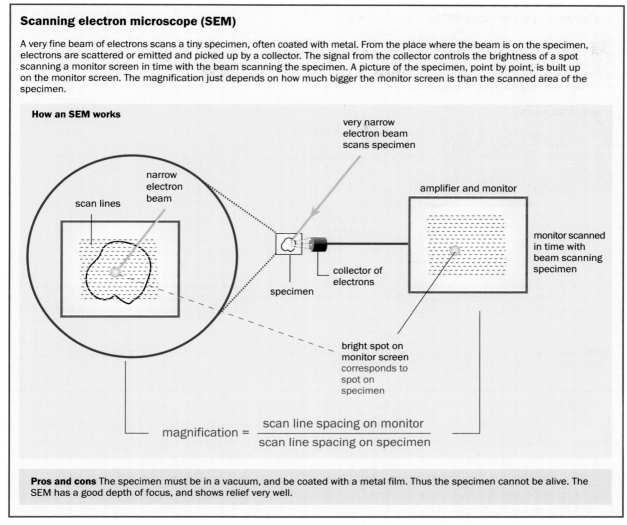

Scanning electron microscope (SEM)

A very fine beam of electrons scans a tiny specimen, often coated with metal. From the place where the beam is on the specimen, electrons are scattered or emitted and picked up by a collector. The signal from the collector controls the brightness of a spot scanning a monitor screen in time with the beam scanning the specimen. A picture of the specimen, point by point, is built up on the monitor screen. The magnification just depends on how much bigger the monitor screen is than the scanned area of the specimen.

How an SEM works

very narrow
electron beam
scans specimen

narrow
electron
beam

scan lines

amplifier and monitor

monitor scanned
in time with
beam scanning
specimen

collector of
electrons

specimen

bright spot on
monitor screen
corresponds to
spot on
specimen

$$\text{magnification} = \frac{\text{scan line spacing on monitor}}{\text{scan line spacing on specimen}}$$

Pros and cons The specimen must be in a vacuum, and be coated with a metal film. Thus the specimen cannot be alive. The SEM has a good depth of focus, and shows relief very well.

The pictures opposite show the images which microscopes can provide, and some of the aspects of material structures which they have revealed.

Order, order

Both the atomic force microscope (AFM) and the scanning tunnelling microscope (STM) allow us to 'see' the atoms of which a material is made. They are both called microscopes, but they don't allow us to look directly at the sample like an optical microscope. Instead, they scan across the surface of the sample, building up an image on the screen of a monitor.

These microscopes tell us about the arrangement of atoms on the surface of a material. Beware! The atoms *inside* the material may be arranged differently. The surface atoms often rearrange themselves because, unlike the atoms inside, they are not surrounded on all sides by other atoms. Other techniques, such as x-ray crystallography, may tell us more about the arrangement of atoms inside the material.

The AFM and the STM can show us the regular, orderly arrangement of the atoms which make up a metal. These materials are crystalline, though they usually don't look like crystals.

In fact, for centuries people have supposed that the atoms in a metal are arranged in a regular, crystalline array. There's a story about this. Robert Boyle (of Boyle's Law) was on the quayside, watching the loading of warships. He saw the neat stacks of cannonballs, and noted how their arrangement gave rise to planes and angles just like those he had seen on the surfaces of crystals. Could crystals be made from millions of tiny, hard spherical particles?

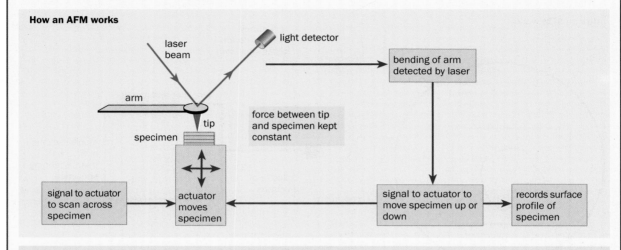

Atomic force microscope (AFM)

Like an old fashioned record player, an atomic force microscope detects the shape of a surface by moving a needle over it. But it detects surface roughness down to the scale of atoms. A fine point is mounted on a cantilever arm (see photograph on page 29). Forces between tip and surface bend the cantilever. A laser beam reflected from the cantilever detects the bending. In one mode of operation, the specimen is moved to keep the force on the tip constant. Up and down movements of the specimen as it is scanned under the tip correspond to its surface profile.

How an AFM works

laser beam

light detector

bending of arm detected by laser

arm

tip

specimen

force between tip and specimen kept constant

signal to actuator to scan across specimen

actuator moves specimen

signal to actuator to move specimen up or down

records surface profile of specimen

Pros and cons The AFM does not need the surface to be specially prepared. The specimen need not conduct electricity. The specimen need not be in a vacuum. The AFM is very suitable for biological work.

This story may be apocryphal, but it does illustrate how people can take an observation and use it to jump to a new idea in some apparently unrelated area. Today, three centuries later, there are new ways of looking at atoms and seeing their arrangements which are, for many materials, just as Boyle suggested.

Measurements show that the smallest atoms are about 0.1 nm in diameter, the largest about 0.5 nm ($1 \text{ nm} = 10^{-9}$ m). So, for the largest atoms, about 2 000 000 will fit side-by-side in just one millimetre.

Try this: Put some marbles in a shallow dish or saucer. (If you don't have marbles, try lentils or dried peas.) Look at the regular pattern they make. Now try grains of rice, short grain and long grain. What sorts of patterns do you see? Marbles are like the spherical atoms of which metals are made; rice grains are more like the elongated molecules of a liquid crystal. Polymers are made of flexible, long-chain molecules. How could you model polymers in the kitchen?

The models opposite show some realistic features of the arrangements of atoms in metals. But, like any model, the comparison has its limitations. There is no attractive force between the ball bearings: tip the dish and they roll apart. And both models are simply two-dimensional representations of a three-dimensional reality.

Modern microscopy allows us to see images of atoms. The images we see depend on the techniques we use to make them. A skilled microscopist must know how to interpret the images, and how to allow for the limitations of the technique.

Crystals you can see, and crystals you can't

Solids form when a liquid cools. The internal structure of the resulting solid may be amorphous (disordered) or crystalline. Rapid cooling tends to trap the particles in an amorphous state, which resembles the disordered arrangement in a liquid. Glass is an example of an amorphous material. On the other hand, slow, controlled cooling can produce a single crystal, which is a highly ordered array of atoms. An important example of a single crystal material is high purity silicon, used for making microchips. A single crystal may weigh several kilograms and be made up of 10^{27} atoms arranged in an almost perfect array.

Models of metals

We can picture a metal as a regular array of spherical atoms, held together by the attractive forces between them. Here are two ways of modelling this in the lab:

Ball bearings in a dish

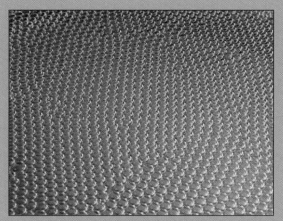

This model can show defects in the lattice: vacancies, where an atom is missing, and interstitials, where a larger atom has been inserted.

Rafts of soap bubbles

Burst a bubble and this model shows a vacancy. It also shows dislocations well; you can even watch them move as planes of bubbles slide past one another.

Many materials, however, are neither completely amorphous nor perfectly crystalline in the solid state; rather, they are polycrystalline. As a liquid cools, crystals start to form at different points within it. Each crystal grows out into the liquid, until it runs into its neighbours. The result is a patchwork of tiny crystals or grains. Where grains meet, the interface is known as a grain boundary. Therefore a polycrystalline material consists of a number of grains, within which the structure is ordered. However, the grains are all oriented differently relative to one another.

The polycrystalline structure of this brass is only seen with a light microscope.

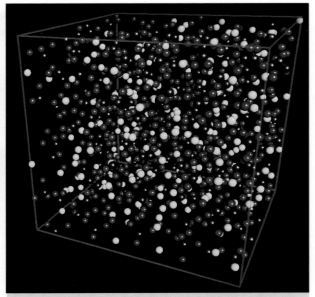

This model shows how atoms may be arranged in one type of glass—sodium disilicate glass. This is an amorphous structure. The positions of the atoms were calculated by computer so that they reproduce the results of detailed x-ray studies of the material.

Try these

1   **Give an example of an amorphous material.**

2   **Give an example of a polycrystalline material.**

3   **Give a single word which is the opposite of 'amorphous'.**

4   **An atom of iron is about 0.25 nm in diameter. How many fit across the point of a steel pin, 0.25 mm in diameter?**

5   **How many iron atoms would fit into a metre length?**

6   **In a scanning electron microscope, suppose that the monitor screen has dimensions 200 mm by 200 mm, scanned in 500 lines. If the corresponding area of the specimen has dimensions 1 μm by 1 μm, what is the linear magnification? What spacing on the specimen can the line scan just resolve?**

7   **How many nearest neighbours does an 'atom' have in the ball bearing model shown on page 107? Is the answer the same for the soap bubble model on the same page?**

Answers 1. glass **2.** most metals **3.** crystalline **4.** 10^6 **5.** 4×10^9 **6.** 200 000, greater than 2 nm **7.** six, yes

YOU HAVE LEARNED

- To understand the properties of a material, we need to know about its structure.

- Structures may be classified as crystalline, polycrystalline and amorphous.

light microscope, crystalline, polycrystalline, amorphous, electron microscope

- The structure of a material may be visible to the naked eye, or we may have to use microscopy or other techniques to reveal it.

- Various techniques are available to show up the structure of materials at the atomic level, e.g. atomic force and scanning tunnelling microscopy.

5.2 A clear view

The eyes of vertebrates (including humans) have many things in common with video cameras. Light from objects far and near is brought to a focus by a lens on a light-sensitive material behind it. In a video camera, a light-sensitive chip records the image; in an eye it is the light receptor cells in the retina. But there is one crucial difference. The lens of a camera is made from glass and, unless you scratch it, it does not lose its transparency over time. The eye lens is composed of a gel containing proteins, called crystallins, arranged in a regular array. Like all living materials, it continually renews itself and it changes with time. Protein molecules are particularly susceptible to oxidative damage (reaction with oxygen) as people age. This tends to decrease the solubility of the crystallins and degrades the crystalline structure of the lens. The result is a loss of transparency and your eyesight gets worse. The effect is not unlike cooking egg white, a clear liquid when raw but white and solid when cooked.

The opacity of the human lens is known as a cataract, and it is most common in people over 70—although it can occur at any age, particularly where other disease such as diabetes is present. At first, a person with a cataract will just need strong spectacles to aid their failing vision. However, once the lens becomes completely opaque it has to be removed, and replaced with an artificial polymer lens. This operation is usually extremely successful.

To understand why a material is transparent or opaque, shiny, clear or coloured, we need to understand what happens when light interacts with it. It's a question of reflection, refraction and absorption. We can understand why cataracts make lenses opaque by thinking about the first two of these.

The 'frosty view' of a cataract sufferer is explained by the breaking down of the uniform crystalline structure of the lens in their eye. But

---

See-through or not?

Explanations involving refraction and reflection

Frosted glass
This is translucent. It lets most light through, but you cannot see things clearly through it.

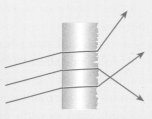

light is refracted at all angles by the rough surface

Car window
This is transparent until shattered in an accident. Then it appears white and opaque.

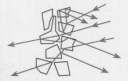

light is reflected and refracted many times by tiny glass fragments, emerging in many directions

Polymers may be:

transparent

or

opaque

amorphous regions

crystalline regions

In polythene, light is reflected and refracted at boundaries between crystalline and amorphous regions, which have different refractive indices. Perspex, which is all amorphous, is transparent.

Frequent reflections and refractions make things translucent or opaque

---

The view of someone with cataracts.

what about coloured glass? And why do metals look shiny, so different from other materials? Now you need to think about how light is absorbed by materials.

Here, electrons are important. The electrons in a material can absorb light; in so doing, they gain energy. They may lose the energy by re-emitting it.

Stiff stuff, tough stuff

Glass is notable for two properties—it is transparent (as discussed above) and it is brittle—it shatters if you hit it hard enough. The structure of glass explains why it is brittle. Glass—and there are many types of glass—consists of a certain amount of silica together with various metal oxides. For example, soda glass—the stuff that laboratory glass tubing is made of—contains about 70% silica with a mixture of sodium oxide and calcium oxide. In fact, the two metal oxides give themselves away when you heat a piece of soda glass; note the orangey-yellow colour of the flame. In the flame test for metal ions, sodium is yellow and calcium is brick red.

Glasses are amorphous, or non-crystalline materials. In the solid state they do not differ very much from the liquid state—there is little order in the way the atoms are arranged—look back to the picture on page 107. The atoms do not arrange themselves in the regular arrays which are typical of crystalline materials such as metals.

It is this underlying difference in internal structure which explains why glass is brittle and copper is not. In the 1920s, Alan Griffith was working at the Royal Aircraft Establishment at Farnborough when he had an important idea. Calculations showed that most materials ought to be ten to a thousand times stronger than is actually observed. Griffith argued that the presence of minute cracks and flaws in a material accounted for this discrepancy. They act as stress raisers; in other words, the stress around a crack can be hundreds or even thousands of times greater than the actual applied stress. Bending the glass slightly makes the crack open up and spread through the material. This is how glass is often cut: the glass merchant scratches a fine line—tap—the sheet of glass falls apart exactly along the line.

Griffith's theory is supported by the superior strength of freshly drawn glass fibres, which have

See-through or not
Explanations involving absorption

Colour in glass

Glass is not completely colourless. If you look through a sheet of glass edge-on, it looks green or blue.

Metal ions in the glass absorb light of certain wavelengths, colouring the glass. Iron makes it green. Copper makes it blue.

Shiny metals

Light penetrates almost no distance into metals. It bounces back from the surface.

Free electrons in the metal absorb the light energy, and then immediately re-emit it. Metals are shiny for the same reason that they conduct electricity; they contain free electrons.

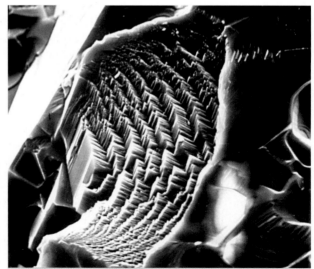

This electron micrograph shows how a crack has moved through a ceramic material—alpha alumina, a form of aluminium oxide. The crack moved diagonally downwards from right to left. It found it easier to travel along horizontal and vertical planes of atoms, and so it has formed a series of steps.

few flaws. In a metal, dislocations in the ordered crystalline structure can help to even out high stress concentrations around a crack.

We can think about how glass breaks in terms of the atoms of which it is made, and the work which must be done in breaking them apart.

- When you bend the glass, it becomes strained elastically. It stores **strain energy**. (This is like the energy stored in a stretched spring.)
- At the tip of the crack, two neighbouring atoms are pulled apart. Work is done in breaking the bond between the atoms.
- Then the next two atoms are pulled apart, and the next two, and so on. The crack moves through the material like a zip being undone. The energy required to do this is known as **fracture energy**.
- Once the glass is broken, it is no longer strained, so it no longer stores strain energy. The energy has been used in breaking the bonds, in sending fragments of glass flying, and in making atoms vibrate.

Strength and toughness

Alan Griffith's work on cracks helped to explain the difference between a brittle material like glass and a tough one like steel. It doesn't take much energy to break glass; it takes a lot to break steel. This is illustrated by the figures in the table.

Material	Approximate fracture energy/$J\,m^{-2}$	Approximate tensile strength/MPa
glass, pottery	1–10	170
cement, brick, stone	3–40	4
polyester and epoxy resins	100	50
nylon, polythene	1000	150–600
bones, teeth	1000	200
wood	10 000	100
mild steel	100 000–1 000 000	400
high tensile steel	10 000	1000

Very approximate figures for the fracture energy and tensile strengths of some common materials. Adapted from J E Gordon *Structures* (1991) Penguin.

The toughness of a material is measured by the energy needed to deepen and extend cracks, creating new fractured surface in the cracks. If the energy available from the stresses in the material is larger than the energy to extend a crack, the crack will propagate and the material will fail. This is why glass can break almost explosively.

The energy used in fracturing a specimen varies a good deal—the new surface may vary in roughness, the number of small flying pieces can alter, and so on. That is why the figures in the table are labelled 'approximate', and are given in very round figures. They do give some rough indication of toughness, however, even though toughness is defined slightly differently, as above. You will see from the table that toughness is not at all the same as tensile strength. Glass has quite a high tensile strength but a low toughness: you have to pull hard to break it but it is brittle and snaps without using much energy. A polyester resin is around ten times tougher than glass, but it breaks at a stress about three times smaller. You don't have to pull so hard to break it, but it doesn't come apart so readily. It is not so brittle. Materials, then, can be tough and strong, tough but weak, brittle and strong or brittle and weak.

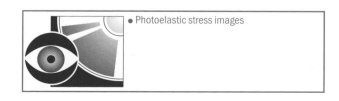

● Photoelastic stress images

Cracks and stress

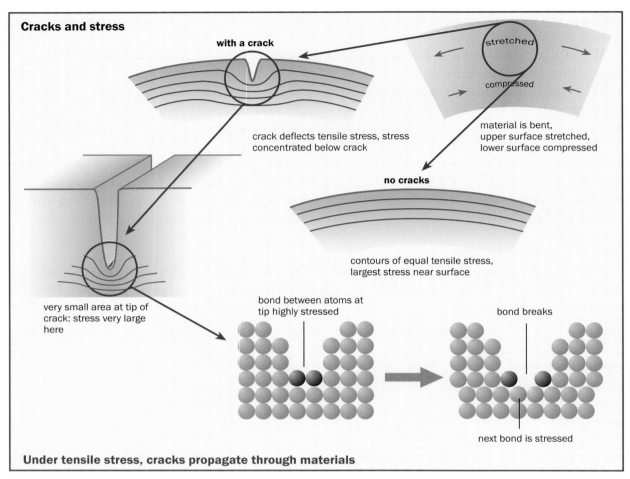

with a crack

stretched

compressed

crack deflects tensile stress, stress concentrated below crack

material is bent, upper surface stretched, lower surface compressed

no cracks

contours of equal tensile stress, largest stress near surface

very small area at tip of crack: stress very large here

bond between atoms at tip highly stressed

bond breaks

next bond is stressed

Under tensile stress, cracks propagate through materials

Stopping cracks propagating

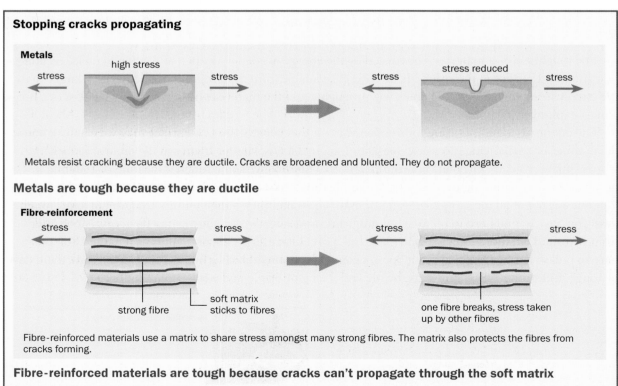

Metals

high stress

stress

stress

stress

stress reduced

stress

Metals resist cracking because they are ductile. Cracks are broadened and blunted. They do not propagate.

Metals are tough because they are ductile

Fibre-reinforcement

stress

stress

stress

stress

strong fibre

soft matrix sticks to fibres

one fibre breaks, stress taken up by other fibres

Fibre-reinforced materials use a matrix to share stress amongst many strong fibres. The matrix also protects the fibres from cracks forming.

Fibre-reinforced materials are tough because cracks can't propagate through the soft matrix

Fracture energy and tensile strength

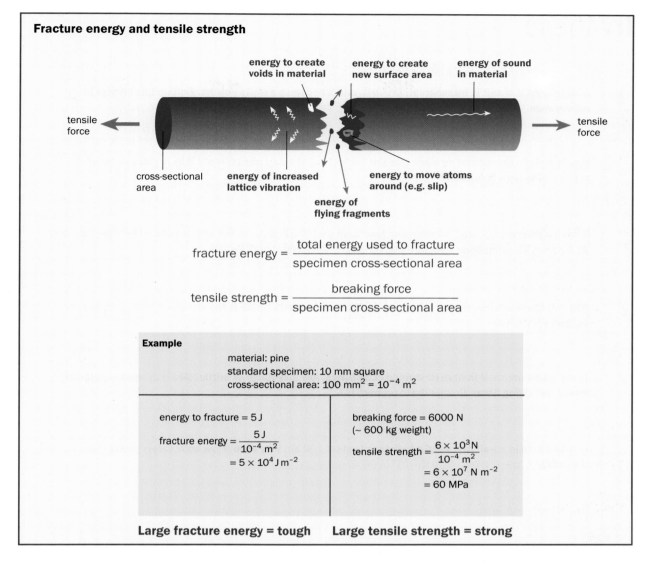

energy to create voids in material

energy to create new surface area

energy of sound in material

tensile force

cross-sectional area

energy of increased lattice vibration

energy to move atoms around (e.g. slip)

energy of flying fragments

tensile force

$$\text{fracture energy} = \frac{\text{total energy used to fracture}}{\text{specimen cross-sectional area}}$$

$$\text{tensile strength} = \frac{\text{breaking force}}{\text{specimen cross-sectional area}}$$

Example

material: pine
standard specimen: 10 mm square
cross-sectional area: $100\ \text{mm}^2 = 10^{-4}\ \text{m}^2$

energy to fracture = 5 J

$$\text{fracture energy} = \frac{5\,\text{J}}{10^{-4}\ \text{m}^2}$$
$$= 5 \times 10^4\ \text{J m}^{-2}$$

breaking force = 6000 N
(~ 600 kg weight)

$$\text{tensile strength} = \frac{6 \times 10^3\,\text{N}}{10^{-4}\ \text{m}^2}$$
$$= 6 \times 10^7\ \text{N m}^{-2}$$
$$= 60\ \text{MPa}$$

Large fracture energy = tough   **Large tensile strength = strong**

Strong and tough

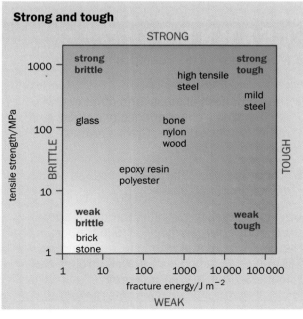

STRONG

tensile strength/MPa

1000 — strong brittle   strong tough

high tensile steel

mild steel

BRITTLE

100 — glass   bone nylon wood

epoxy resin polyester

10 —

TOUGH

weak brittle   weak tough

1 — brick stone

fracture energy/J m^{-2}

1   10   100   1000   10 000 100 000

WEAK

Imagine breaking identical rods of glass and steel. The tensile strength of glass is roughly half that of mild steel. They can withstand similar forces. But the fracture energy of glass is much, much less than that of mild steel, at least 10 000 times less. So glass is strong but brittle; mild steel is strong and tough. Brick and stone, by contrast, are both weak and brittle: they break at a low tensile stress and come apart without using much energy to do so. You might like to think of a possible weak but tough material—fibre-reinforced jelly, perhaps? There are few real examples, however.

● Materials database

Try these

Questions 4–6 relate to the data in the table on page 111.

1 A large crystal of salt is transparent. Crush it, and it becomes a white powder. Explain this change in appearance.

2 How is a glass shower screen made translucent instead of being transparent? Why is glass stronger in compression than in tension?

3 Some polythene bags used for wrapping food have a whiteish, translucent appearance while others are clear. Which have an amorphous structure and which are semicrystalline? Explain.

4 Which is the tougher material, mild steel or high tensile steel? Which is the stronger? Give some data to support your answer.

5 Epoxy resins are used in fibre-reinforced composites. Are they tough or brittle? Are they weak or strong? How does adding fibres change these properties?

6 To drill a hole in a wall, you need a high tensile steel drill bit. Why should you wear safety glasses when drilling the hole?

YOU HAVE LEARNED

- The appearance of a material depends on how light is reflected and refracted at the material's surfaces.

- Its appearance also depends on how it absorbs light which falls on it.

- Light may be absorbed by electrons within a material.

- Glasses have a disordered, amorphous structure.

- The brittleness of glass (and other ceramic materials) is a consequence of defects such as fine surface cracks. Cracks propagate easily through glass.

- Metals are generally tough, because cracks do not propagate easily.

- When a crack propagates through a brittle material, energy is needed to tear apart atoms, create new surfaces, move atoms around and send shock waves through the material. This is fracture energy.

 reflection, refraction, refractive index, transparent, absorption, fracture energy, tensile strength, yield stress

5.3 Making more of materials

Gold is a beautiful material. For thousands of years it has been used for making attractive objects—bracelets, medallions, rings, brooches. Today, these may seem to be purely decorative items, but to the wearer they were enormously useful, as symbols of power, status and wealth.

Gold was one of the first metals to be worked by people. It is one of the few metals which can be found in its native state, that is, as a pure metal which does not have to be extracted from an ore. Flakes of unoxidised gold could be found in riverbeds, and gold nuggets could be extracted from exposed seams in rocks.

There are several ways of working gold into a desired shape. The earliest gold artefacts were made simply by hammering out the metal with hand-held stones. The resulting sheet could then be scratched or pierced to produce the desired effect. A more

thick. Gold leaf has been widely used for centuries (and still is today), for example in religious icons, especially in the Orthodox Church, and for decorating public buildings such as the domes of mosques.

More elaborate, three-dimensional shapes can be produced by melting the gold and pouring it into a mould. (Gold melts at a little over 1000 °C.) The mould itself is often made by the 'lost wax' process. The object is modelled in wax, and then wet sand is packed around it. The wax is melted and poured out, leaving a hollow mould into which molten gold is poured. This technique can produce very finely detailed objects.

A bit of a mouthful

Dentists use about 60 tonnes of gold each year, worldwide. Gold is useful for fillings because it does not corrode or tarnish in the damp environment of the mouth, so you can be proud of your shiny golden smile.

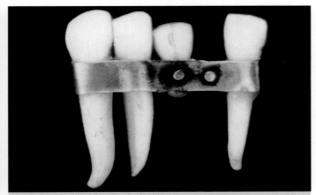

This is the work of an Etruscan dentist of 2500 years ago. A band of gold around three of the patient's own teeth is used to hold in place a false tooth, made from a calf's tooth.

This gold mask was made in the 17th century by the 'lost wax' process. It comes from the Ashanti region of present-day Ghana, where West African metal-working techniques had become highly sophisticated.

Fillings have to be able to withstand the crushing forces when your jaws close on your food—up to 400 N between the back teeth. So the gold used by dentists is an alloy of gold with other metals such as silver, palladium and platinum. Alloying increases the hardness of the metal.

Gold is a ductile metal. It can be hammered, rolled or drawn out into a desired shape, without having to melt it. Ductility is a typical property of metals which makes them different from ceramics and polymers. Think what happens if you try to hammer glass or plastic.

There are mismatches called dislocations in the tidy rows of atoms. It is the movement of these

recently developed technique is rolling, which makes use of the ductility of gold. The metal is passed between a pair of rollers. It is gradually squeezed thinner and thinner; the final gold leaf produced may be only a few hundreds of atoms

Shaping and slipping

Atoms in gold are in a regular array: a crystal lattice. To shape the metal, one layer must be made to slide over another.

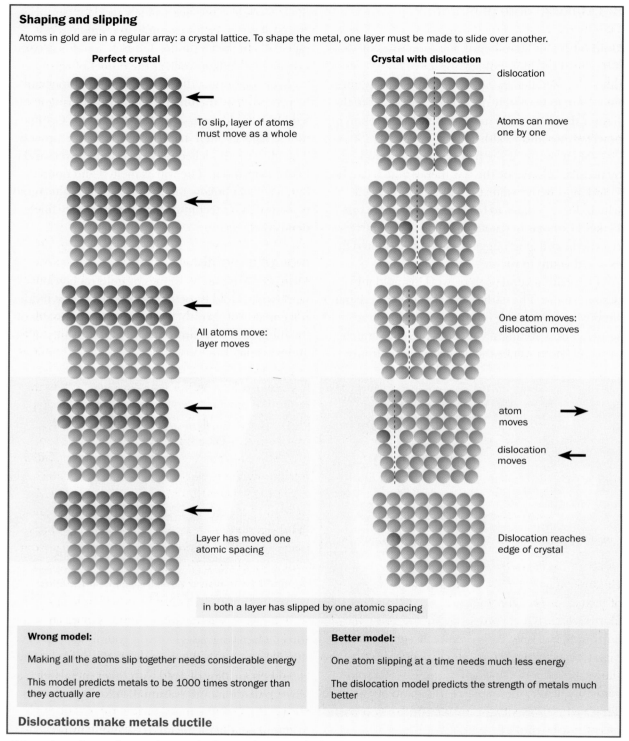

Perfect crystal

To slip, layer of atoms must move as a whole

All atoms move: layer moves

Layer has moved one atomic spacing

Crystal with dislocation

dislocation

Atoms can move one by one

One atom moves: dislocation moves

atom moves

dislocation moves

Dislocation reaches edge of crystal

in both a layer has slipped by one atomic spacing

Wrong model:

Making all the atoms slip together needs considerable energy

This model predicts metals to be 1000 times stronger than they actually are

Better model:

One atom slipping at a time needs much less energy

The dislocation model predicts the strength of metals much better

Dislocations make metals ductile

dislocations through metal crystals that makes metals ductile. This is a good thing if you are trying to shape a piece of metal. But it is a disadvantage if you want to make a very strong object out of metal. Because of dislocations, metals may be as much as 1000 times weaker than they would otherwise be.

Dislocations can work the other way round too. Flex a copper wire or a paper clip repeatedly between your fingers. This can create dislocations. When there are many of them, they get entangled and stop each other moving. The metal is now *less* ductile. It has been 'work hardened'.

Pure metals and alloys

Pure metals tend to be very soft. Soft metals may gradually deform under their own weight, like the lead which slides down church roofs over the years; this is the phenomenon of creep. Alloying introduces atoms of other elements, usually of different sizes. This disrupts the regularity of the crystal structure and makes it harder for atoms to slide past one another.

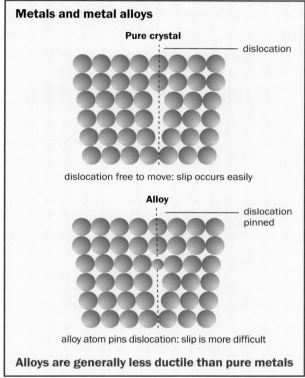

Metals and metal alloys

Pure crystal

— dislocation

dislocation free to move: slip occurs easily

Alloy

— dislocation pinned

alloy atom pins dislocation: slip is more difficult

Alloys are generally less ductile than pure metals

Ceramics, familiar and unfamiliar

Ceramic materials have been around for thousands of years. Pottery is an ancient example, and it illustrates the typical properties of ceramics. It is hard, stiff and wear resistant. It can survive at fairly high temperatures and resists corrosion. Pottery also shows the main weakness of ceramics—they are generally brittle, so they can't withstand impacts. That's why archaeologists usually find only broken bits of ancient pots.

Other traditional ceramic materials include brick and concrete, and there are many natural materials which we could classify as ceramics, such as flint and stones. Ceramics are crystal structures, generally built from ions of two or more elements.

Increasingly, engineers are turning to newer ceramics. Pottery is made from clay, which can have a very variable composition, but engineering ceramics are made from highly purified starting materials, so that their composition and structure can be carefully controlled. From the names of some of these materials—aluminium oxide, silicon nitride, silicon carbide—you will see that they are relatively simple compounds in which the atoms are strongly bonded together.

Fatal flaws

Besides brittleness, there is another big problem with ceramics: they are difficult to shape. It's easy to understand why this is: it's because of their hardness. They can't be melted and shaped, like most metals or polymers. So ceramic objects are often made by shaping the raw materials and then heating them. Think about how a cup is made. Clay is shaped, and then fired in a kiln. Some of the minerals in the clay melt to form a glassy liquid, and when cooled and solidified this glassy material 'glues' the material together.

Engineering ceramics may be processed in the same way. Silicon powder is moulded to the shape of, say, a turbine blade. Then it is heated with nitrogen, and a chemical reaction forms silicon nitride. An alternative approach is sintering. Silicon nitride powder is placed in a mould and heated under high pressure. The surfaces of the particles melt and flow, so that they become glued together. (This is rather like crushing snowflakes together to make a snowball.) Another way is to sinter powder with a laser beam. Now the object to be made is defined by the instructions for steering the laser beam. Thus the object can be downloaded as software and be made on the spot. No need to stockpile spare parts.

The key to making a good ceramic object is to avoid flaws in its structure. Any tiny cracks, holes or impurities can result in a fatal weakness. Just as with glass, a crack can propagate rapidly through a ceramic material, and this explains why they are generally much weaker in tension than in compression.

It may be possible to have the best of both worlds by combining ceramics and metals. A 'cermet' is a combination of ceramic particles in a metal matrix (see picture on page 104). Such a material can be hard, tough and strong.

Ceramics versus metals

Ceramics have rigid structures

Covalent structures

example: silica (also diamond, carborundum)

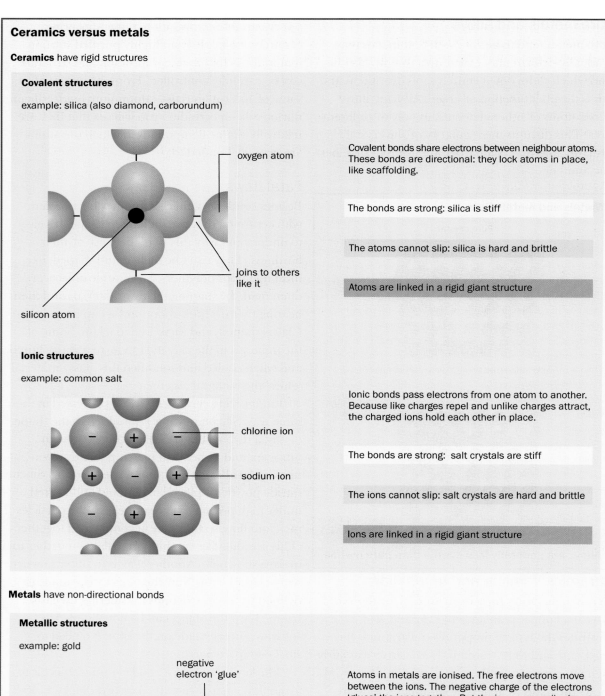

oxygen atom

joins to others like it

silicon atom

Covalent bonds share electrons between neighbour atoms. These bonds are directional: they lock atoms in place, like scaffolding.

The bonds are strong: silica is stiff

The atoms cannot slip: silica is hard and brittle

Atoms are linked in a rigid giant structure

Ionic structures

example: common salt

chlorine ion

sodium ion

Ionic bonds pass electrons from one atom to another. Because like charges repel and unlike charges attract, the charged ions hold each other in place.

The bonds are strong: salt crystals are stiff

The ions cannot slip: salt crystals are hard and brittle

Ions are linked in a rigid giant structure

Metals have non-directional bonds

Metallic structures

example: gold

negative electron 'glue'

gold ion

Atoms in metals are ionised. The free electrons move between the ions. The negative charge of the electrons 'glues' the ions together. But the ions can easily change places.

The bonds are strong: metals are stiff

The ions can slip: metals are ductile and tough

Ions are held together, but can move

Fibres and fabrics, polymers and plastics

Polymers are substances made from long-chain molecules—polythene, nylon, Perspex. These are synthetic polymers, invented in the twentieth century. There are many natural polymers, too—wool, cotton, leather. Many biological molecules are polymers—starch, proteins and, of course, DNA. Polymers tend to be flexible but strong—that's why most of our clothes are made of polymers.

A molecule of polythene—more correctly, poly(ethene)—is a long chain of identical, repeating units. A molecule like this is very floppy, because it is free to rotate about its bonds. However, the atoms are joined together by strong, covalent bonds so they are difficult to break. So polythene is flexible but strong.

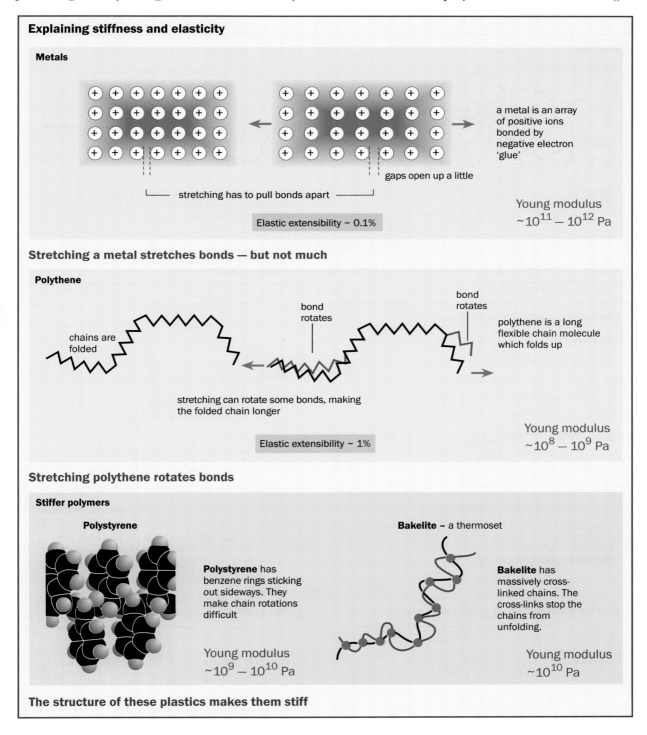

Explaining stiffness and elasticity

Metals

a metal is an array of positive ions bonded by negative electron 'glue'

gaps open up a little

stretching has to pull bonds apart

Elastic extensibility ~ 0.1%

Young modulus
$\sim 10^{11} - 10^{12}$ Pa

Stretching a metal stretches bonds — but not much

Polythene

bond rotates

bond rotates

polythene is a long flexible chain molecule which folds up

chains are folded

stretching can rotate some bonds, making the folded chain longer

Elastic extensibility ~ 1%

Young modulus
$\sim 10^{8} - 10^{9}$ Pa

Stretching polythene rotates bonds

Stiffer polymers

Polystyrene

Bakelite – a thermoset

Polystyrene has benzene rings sticking out sideways. They make chain rotations difficult

Bakelite has massively cross-linked chains. The cross-links stop the chains from unfolding.

Young modulus
$\sim 10^{9} - 10^{10}$ Pa

Young modulus
$\sim 10^{10}$ Pa

The structure of these plastics makes them stiff

Rubber control

Rubber latex is a runny white liquid, the sap of the rubber tree, a native of South America. Christopher Columbus learned from the indigenous people how to make heavy black rubber balls. At first rubber was regarded simply as a curiosity, an amusing, sticky, bouncy substance, useful for waterproofing clothing and hosepipes. However, when Charles Goodyear invented the process of vulcanisation in 1839, it suddenly became a valuable material. To vulcanise rubber, it is heated with sulphur. Cross-links are formed between the polymer chains by the sulphur atoms. The more sulphur you add, the more cross-links form and the stiffer the rubber becomes. Control of the structure of the material gave control over its properties. So rubber found many new applications—in tyres, shock absorbers, electrical insulation.

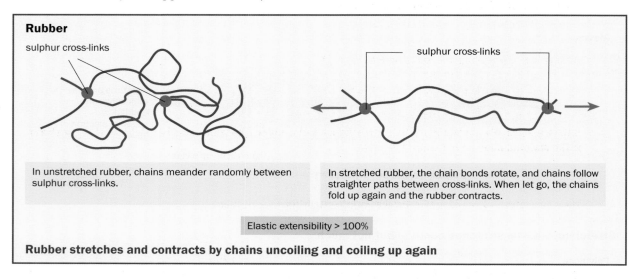

Rubber

sulphur cross-links

sulphur cross-links

In unstretched rubber, chains meander randomly between sulphur cross-links.

In stretched rubber, the chain bonds rotate, and chains follow straighter paths between cross-links. When let go, the chains fold up again and the rubber contracts.

Elastic extensibility > 100%

Rubber stretches and contracts by chains uncoiling and coiling up again

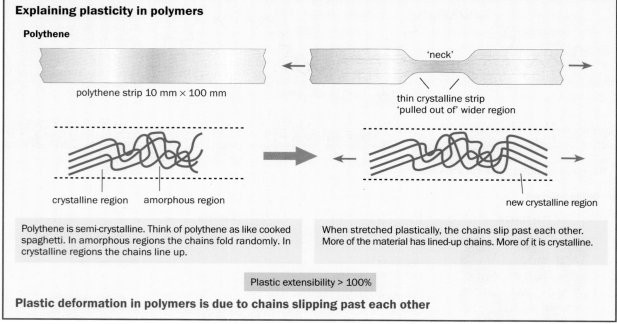

Explaining plasticity in polymers

Polythene

polythene strip 10 mm × 100 mm

'neck'

thin crystalline strip 'pulled out of' wider region

crystalline region   amorphous region

new crystalline region

Polythene is semi-crystalline. Think of polythene as like cooked spaghetti. In amorphous regions the chains fold randomly. In crystalline regions the chains line up.

When stretched plastically, the chains slip past each other. More of the material has lined-up chains. More of it is crystalline.

Plastic extensibility > 100%

Plastic deformation in polymers is due to chains slipping past each other

Silly putty—bounce and flow

Silly putty has some interesting behaviour. Pull it slowly and it stretches and stretches. By pulling slowly, you are allowing the molecules time to slip and slide past one another. Pull it quickly and it snaps. The molecules haven't had time to rearrange themselves. Drop it on the floor and it bounces. Here you are using the material's elasticity. The molecules scrunch up a bit, then spring back to their original positions.

Try these

1 There are three types of strong bond between atoms: covalent, ionic and metallic. Which is/are non-directional? Which bonding is associated with toughness of the material?

2 Metals contain dislocations, which allow easy slipping of layers of atoms. What aesthetic property of metal objects does this fact help to make possible?

3 What is happening when a metal is work-hardened?

4 If the maximum elastic strain of a polymer material is 1%, and its Young modulus is 10^8 Pa, what is the stress in the material when stretched to this extent?

5 What does plastic stretching of a polymer have to do with crystallisation?

6 Rubber bands stretch a lot; car tyre rubber does not. What is the reason for the difference?

Answers **1.** metallic; metallic **2.** ductile metals can be formed into complex and delicate shapes **3.** new dislocations are created which get in one another's way **4.** stress = 10^6 Pa **5.** in stretching, new crystalline regions are formed **6.** the longer or shorter lengths of polymer chain between cross-links

YOU HAVE LEARNED

- Metallic bonds are strong, non-directional bonds; these account for the toughness of metals.

- Because metallic bonds are non-directional, atoms can slide easily past one another; this accounts for the ductility of metals.

- Strong ionic or covalent bonds between the atoms account for the stiffness of ceramic materials.

- Polymers have long-chain molecules of repeating units.

- Cross-linking of chains can make a polymer rigid.

- Rotation and slipping of chains can make a polymer plastic.

amorphous, crystalline, ductile, dislocation, bonding, tough, brittle, elastic, plastic

5.4 Controlling conductivity

Semiconductors are fascinating materials, and the technology which uses them has transformed our world. They are electrical conductors, but their conductivity is much less than that of metals. Silicon and germanium, two examples of semiconductors, lie in the middle of the Periodic Table between the metals on the left and the insulators on the right. Because of their importance in modern electronic devices, an industry has developed to produce extremely pure semiconducting materials, so that silicon crystals are available with 99.999 999 9% purity—just one impurity atom remains among a billion silicon atoms.

Transistors are the semiconductor devices that make possible our personal computers, hi-fi systems, body scanners, aircraft control systems, telephone networks etc etc. Transistors are used to control and amplify electrical signals; the latest microprocessor at the heart of a personal computer contains ten of millions of them, all packaged in a single integrated circuit or chip. The task of producing the detailed design of such a chip is beyond any human, so computers design each new generation of chips according to instructions laid down by a team of engineers—chips designing chips.

Birth of the information age

The first transistor was demonstrated in 1947. It was invented by three physicist-engineers working

The first transistor was a ramshackle affair, made from a small piece of germanium. The 'business end' is the point where the triangle of plastic presses a split piece of gold foil down on to the horizontal slab of germanium. Today, many millions of transistors are integrated into a single semiconductor chip.

● How resistivity changes with temperature

The full story of the invention of the transistor is told in Michael Riordan and Lillian Hoddeson's book *Crystal Fire: The Birth of the Information Age.* Here they describe the impact of the transistor:

"*Nations that have embraced the new information technologies based on transistors and microchips have flourished. Japan and its retinue of developing eastern Asian countries increasingly set the world's communication standards, manufacturing much of the necessary equipment. Television signals penetrate an ever growing fraction of the globe via satellite. Banks exchange money via rivers of ones and zeroes flashing through electronic networks around the world. And boy meets girl over the Internet.*

"*No doubt the birth of a revolutionary artefact often goes unnoticed amid the clamour of daily events. In half a century's time, the transistor, whose modest role is to amplify electrical signals, has redefined the meaning of power, which today is based as much on the control and exchange of information as it is on iron or oil. The throbbing heart of this sweeping global transformation is the tiny solid-state amplifier invented by Bardeen, Brattain and Shockley. The crystal fire they ignited during those anxious postwar years has radically reshaped the world and the way its inhabitants now go about their daily lives.*"

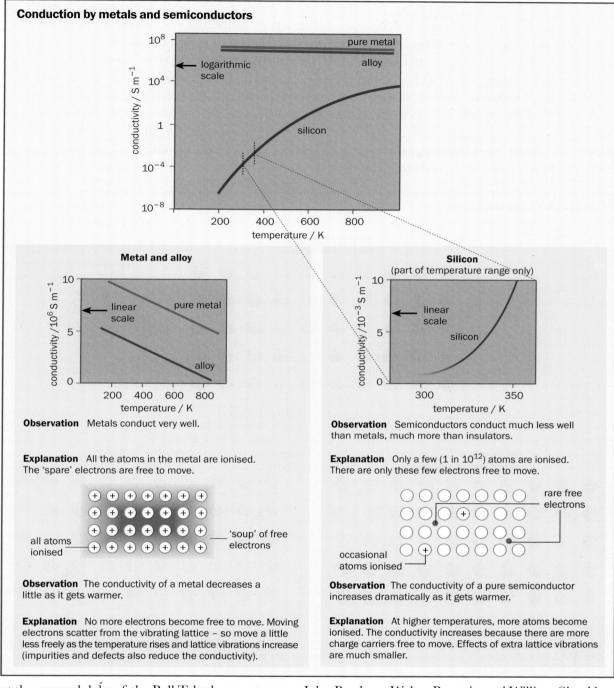

Conduction by metals and semiconductors

Metal and alloy

Observation Metals conduct very well.

Explanation All the atoms in the metal are ionised. The 'spare' electrons are free to move.

Observation The conductivity of a metal decreases a little as it gets warmer.

Explanation No more electrons become free to move. Moving electrons scatter from the vibrating lattice – so move a little less freely as the temperature rises and lattice vibrations increase (impurities and defects also reduce the conductivity).

Silicon
(part of temperature range only)

Observation Semiconductors conduct much less well than metals, much more than insulators.

Explanation Only a few (1 in 10^{12}) atoms are ionised. There are only these few electrons free to move.

Observation The conductivity of a pure semiconductor increases dramatically as it gets warmer.

Explanation At higher temperatures, more atoms become ionised. The conductivity increases because there are more charge carriers free to move. Effects of extra lattice vibrations are much smaller.

at the research labs of the Bell Telephone company: John Bardeen, Walter Brattain and William Shockley. Their device amplified a small electric current by a factor of almost one hundred. Until this time, electrical amplification was done using valves, glass vacuum tubes through which electric currents flow. Valves still have their uses today, but they are bulky and take time to warm up. The attraction of the transistor is that it is a solid-state device—it is entirely made from solid materials—and so it is much more robust and reliable.

A metal is a worse conductor when it is heated, because the vibrating ions scatter the moving electrons. This is true for semiconductors too, but it is masked by a much bigger effect, the great increase in the number of free electrons as they break away from their atoms.

Objects made of materials that you generally think of as insulators will often conduct a tiny current. This may be because of moisture on the surface: not really a property of the material, but important, especially with glass insulators. Making an insulating material hot will often make it conduct a little better, by freeing charge carriers to move. For example, glass conducts quite well when it is red-hot. Sodium or other ions in the glass can get enough energy to hop about in the amorphous structure and carry a current.

Properties of materials come in packages, because the structure of each affects a number of different properties. For example, the free electrons in a metal (a) glue the ions together with non-directional bonds making a metal strong but ductile, (b) allow the metal to conduct electricity well, and (c) make the metal shiny by absorbing and re-emitting light striking the surface. Three seemingly distinct properties all have the same origin.

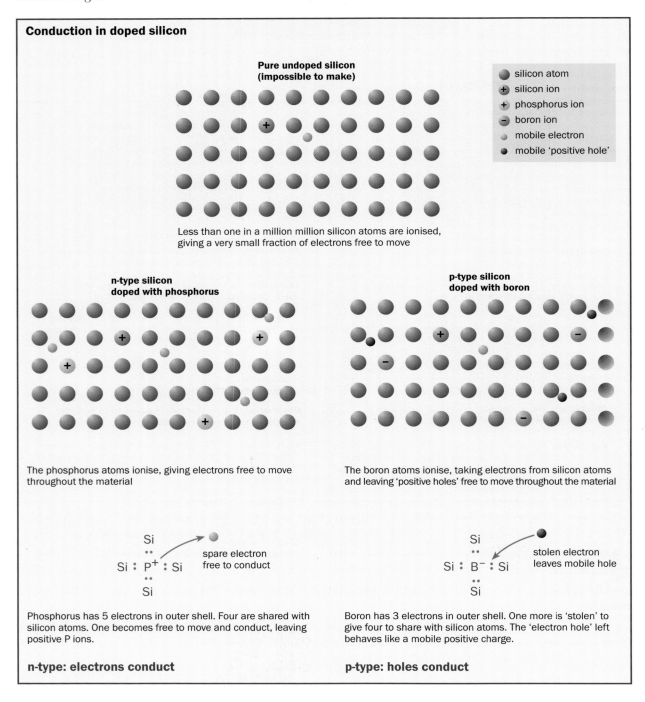

Conduction in doped silicon

Pure undoped silicon (impossible to make)

- silicon atom
- silicon ion
- phosphorus ion
- boron ion
- mobile electron
- mobile 'positive hole'

Less than one in a million million silicon atoms are ionised, giving a very small fraction of electrons free to move

n-type silicon doped with phosphorus

The phosphorus atoms ionise, giving electrons free to move throughout the material

Si : P$^+$: Si → spare electron free to conduct

Phosphorus has 5 electrons in outer shell. Four are shared with silicon atoms. One becomes free to move and conduct, leaving positive P ions.

n-type: electrons conduct

p-type silicon doped with boron

The boron atoms ionise, taking electrons from silicon atoms and leaving 'positive holes' free to move throughout the material

Si : B$^-$: Si → stolen electron leaves mobile hole

Boron has 3 electrons in outer shell. One more is 'stolen' to give four to share with silicon atoms. The 'electron hole' left behaves like a mobile positive charge.

p-type: holes conduct

Doping and devices

Integrated circuits are made in 'clean rooms', where the air is purified to remove almost every speck of dust. This is essential, because to make a chip you have to control the purity of the semiconducting material. Any impurities will undo all your hard work.

A chip is a miniaturised electric circuit, with insulating, conducting and semiconducting regions. Making a chip relies on being able to control the conductivity of silicon. To make an insulating region, it can be exposed to oxygen or nitrogen to produce silicon oxide or nitride. To produce a conducting region, it may be coated with gold. But the vital step which produces the semiconducting parts of the transistors is the process called doping. Ions are sprayed onto the exposed surface of the silicon and become absorbed into its crystalline structure. Depending on the type of ions used, this produces either n-type or p-type silicon.

- In n-type, the impurities provide extra electrons which increase the material's conductivity.
- In p-type, the impurities provide 'holes' which behave like positively charged conduction electrons.
  A current flows when holes move from positive to negative.

See chapter 2 for uses of microcircuits in sensors.

Properties go together

Metals

because
the free electrons in metals:

conduct well	are mobile and so carry electric current
are shiny	oscillate in light, scattering light photons
are stiff	'glue' ions together strongly
are ductile	provide a non-directional 'glue', letting ions slip

Ceramics

because
the ionic or covalent bonds holding them together:

are insulators	lock electrons to ions or atoms, so none are free to move
are stiff	are strong bonds, hard to stretch
are brittle	are directional bonds, so that atoms or ions cannot slip

Polymers

because
the covalent bonds stringing monomers in long chains:

are insulators	lock electrons to atoms, with none free to move
are often flexible	can rotate, letting chains stretch or fold
are often plastic	make chains which can slip past one another

Ceramics, glasses and polymers

because
there are no free electrons to scatter light photons,

either may be transparent or may be opaque/ translucent	letting light go straight through if the material is uniform throughout or refracting and reflecting light many times at boundaries between different regions

The bonding and structure of a material explain whole sets of properties

---

- High-temperature superconductivity
- Materials: Yesterday and tomorrow

Try these

1   Which has the greater conductivity at room temperature, a metal or a semiconductor?

2   What particles are the current-carriers in the following materials: a metal, an undoped semiconductor, a p-type semiconductor, hot glass?

3   For each of the following properties, say whether it is characteristic of a metal or of an insulator: low electrical conductivity; opaque to light; good thermal conductivity; ductile; brittle.

4   Impurity atoms in a metal obstruct the movement of free electrons. What can you say about the resistivity of an impure metal compared to that of a pure metal?

5   Phosphorus is to the right of silicon in the Periodic Table of the elements. Will doping silicon with phosphorus produce a p-type or n-type material?

6   Suppose you are given data on the conductivity of a metal and of a semiconductor over a wide range of temperatures, and are asked to plot conductivity against temperature. Sketch the shapes of the graphs you expect. Would you expect to use linear or logarithmic scales of conductivity?

Answers   **1.** metal   **2.** electrons, electrons, holes, ions   **3.** insulator, metal, metal, metal, insulator   **4.** resistivity greater
5. n-type   **6.** metal: straight line sloping down, linear scale; semiconductor: rising curve, logarithmic scale

YOU HAVE LEARNED

- Semiconductors have electrical conductivity intermediate between that of metals and that of insulators.

- The conductivity of a metal decreases gradually as the temperature increases.

- The conductivity of semiconductors increases rapidly with temperature, as more electrons break free.

- The conductivity of a semiconductor can be controlled by doping with other elements.

- In n-type material, the charge-carriers are electrons; in p-type, they are holes.

semiconductor, doping, free electrons, electrical conductivity, resistivity

Summary checkup

✓ Structures:

- Understanding the structure of a material can help us to understand its properties; by altering the structure we can control the properties.

- Metals generally have an ordered, polycrystalline structure.

- Glasses have an amorphous structure, similar to that of a liquid.

✓ Bonding:

- Insulators (such as ceramic materials) are generally held together by strong, directional bonds (ionic and covalent).

- Metals are held together by strong, non-directional bonds (metallic bonding).

- Polymer molecules have a long-chain structure, with strong bonds within the chains.

✓ Mechanical properties:

- When a metal or ceramic is stretched elastically, the bonds between neighbouring atoms are extended very slightly.

- When a polymer is stretched elastically, the atoms rotate about their bonds.

- Metals are ductile because of the presence of dislocations which can move through the material.

- Ceramics are brittle because cracks and other flaws limit their strength.

✓ Optical properties:

- Metals are opaque and shiny because their conduction electrons scatter light back (they absorb and re-emit the light).

- Transparent insulators like glass and Perspex have no free electrons to scatter light back again.

- Opaque or translucent insulators such as polythene have a microscopic structure in which light is repeatedly scattered at boundaries between regions in the material.

✓ Electrical properties:

- Metals are good conductors; ceramics and polymers are generally good insulators.

- Semiconductors have intermediate conductivity.

- The conductivity of metals decreases gradually with increasing temperature, while the conductivity of semiconductors increases rapidly.

- The conductivity of a semiconductor can be increased by doping with atoms of other elements nearby in the Periodic Table.

Questions

1 Materials may be grouped or classified in a variety of ways. Here are some examples of classes:

- *metals*
- *polymers*
- *ceramics*
- *semiconductors*

For each of the statements which follow, say which one or more of these classes it might relate to.

(a) High electrical conductivity because of the presence of many free electrons.

(b) Electrical conductivity increases rapidly with temperature, as the number of free electrons increases.

(c) Transparent, crystalline materials.

(d) Non-directional bonds result in ductility.

(e) May be capable of large elastic deformation as atoms rotate about bonds.

(f) Suffer from weakness due to the presence of surface cracks.

2 We can predict how strong a material might be, if we know how strong the bonds between neighbouring atoms are. However, most materials turn out to be much weaker than our predictions.

(a) Explain why glass is generally a weak material.

(b) Explain why metals are generally weaker than predicted.

(c) Glass-fibre reinforced plastics are strong materials. Explain how fibre reinforcement increases the strength of plastic.

3 Use the data on page 111 to estimate

(a) the tensile force needed to stretch and break a wooden metre rule

(b) the tensile force needed to break a tooth

(c) the tensile force needed to break a nylon fishing line as thin as a hair

(d) the tensile force needed to break a 5 mm diameter high tensile steel cable.

4 It is possible to break a paper clip by repeatedly flexing the wire backwards and forwards.

(a) Describe what you would feel as you did this.

(b) Explain in terms of the changing internal structure of the wire why this happens.

5 The graph shows how the resistivity of each of three materials changes as the temperature increases. The three materials are:

- *a pure metal*
- *a pure semiconductor*
- *a doped semiconductor*

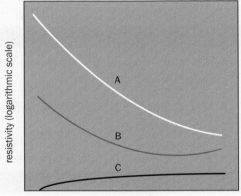

(a) Say which line on the graph corresponds to each material.

(b) Explain your choices.

6 A car windscreen is transparent. However, after a traffic accident, there is often a white pile of fragments of broken windscreen lying in the road.

(a) Explain how the appearance of the windscreen glass changes from colourless and transparent to white.

(b) Windscreens use glass 'toughened' by a heat treatment which leaves the surfaces under tension. Explain how this tension makes them less likely to fracture.

7 Look at the table of strength against fracture energy on page 111, and consult the CD-ROM Materials Database.

(a) How much tougher is mild steel than high tensile steel (an alloy)?

(b) Explain the difference in terms of their internal structures.

(c) For what kinds of uses is mild steel to be preferred to high tensile steel?

6 Wave behaviour

Think of the rainbow reflections from a compact disc or of the colours in a butterfly's wings. Think of a melody played on a solo instrument or the beauty of the song of a blackbird. All have to do with how waves behave. In this chapter we tell the story of how ideas about waves and the nature of light have changed over the centuries. We will look at:

● Colours and sounds produced when waves 'add together'

● Answers to the question, 'what is light?'

● Effects of interference and diffraction

● How to represent waves by spinning arrows called phasors

6.1 Beautiful colours, wonderful sounds

The beautiful shimmering blue wing of the morpho butterfly is an illusion: it is not really blue at all. Seen from many angles it is just a dull brown or grey colour. But, seen at the proper angle, the wing reflects back blue light from the white sunlight falling on it, absorbing the other colours in the light.

The delicate colours of the semi-precious stone opal are made by light scattering from regular arrays of tiny spheres of silica. Look at light reflected by the silvery surface of a compact disc: again you see colours—red, yellow, green, blue. This time the colours come from light scattered at the equally spaced rows of the finely spaced spiral of 'dots' on the disc (chapter 3, page 63). Light reaching your eye from an opal or from a compact disc comes from many rows or lines on the surface, all equally spaced. Light scattered to your eye from one row travels a bit further than the light scattered

The wing of the Brazilian butterfly *Morpho sulkowski*. The wing is covered in microscopic thin transparent platelets, rather like tiles, which are just the right thickness and are laid out on the surface in just the right way to reflect back blue light at certain angles.

Close-up of the surface of an opal. Light is diffracted from regular arrays of spheres of silica in its structure.

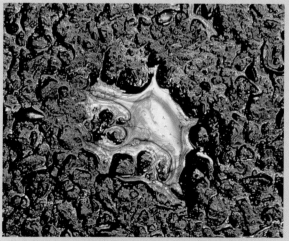

Oil slick on a puddle, with colours from combining light reflected from the front and back of the oil film.

from the one in front, and is delayed. If this time lag brings wave crests and troughs together, light of that colour is weakened; if it brings a crest together with a later crest that colour is strengthened.

Next time it rains, go out and look for oil films on puddles. You can often see a shifting pattern of colours on the surface of these films—especially if the sun is out. This time the colours are produced by light coming to your eye from both front and back of the oil film, with the light from the back having to travel a bit further than the light from the front. Colours are again strengthened or weakened according to whether the time delays bring crests together with crests or with troughs. Soap bubbles show similar colours; again light reflects from both the inner and outer surfaces.

All these colours have something remarkable in common. They change as you move your eye— which is why they often shimmer. That's a way to test whether colours depend on waves coming together with time lags depending on differences in path. Move your head and you change the paths, and so the colours change. Think for example of the iridescent coloured plumage on the neck of a Mallard duck. By contrast, the colour of a shirt or blouse doesn't change if you see it from different angles. Ask yourself, where are these different kinds of colour made—in your eye or in the coloured object, or in neither? It's a subtle question.

These colours come from light combining with light. So how can white light plus white light make blue light—or red, or green? The answer lies in the different paths the light takes. These path differences decide by how much the light waves of a particular colour are in or out of step with one another, and so how they combine when they come together, perhaps in your eye.

Superposition: waves on top of one another

One day in 1933, the crew of the US ship Ramapo crossing the North Pacific were horrified to see a wave over 30 metres high bearing down on their ship. There are many such stories about the sudden appearance of gigantic ocean waves, apparently from nowhere. These rare and terrifying monsters—known to marine insurers as 'hundred-year waves'—arise from the chance meeting of

many waves which just happen to pile on top of one another at a given moment. The opposite can happen too, a moment of calm as waves which are peaking meet waves which are sinking. Crews of ocean racing yachts often wait for such a 'flat' before turning the boat onto a different course.

Waves on water simply pass right through one another. They aren't like moving objects which can hit one another and bounce. They are just movements of the water surface, and the motion from

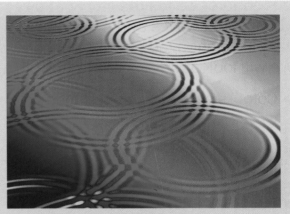

Water ripples moving through each other.

A stormy sea showing little waves on top of big ones.

A famous Japanese print *The Wave*, showing a huge stylised breaker.

one wave just adds onto the motion from another. Like snakes in a snake pit, waves criss-cross one another without noticing that others are there. The same idea lies behind thinking of a complex sound as a spectrum of different frequencies all sounding together, as in chapter 3.

The idea that motions from different waves all pile on top of one another at a given place is called the **principle of superposition**. 'Superposition' just means 'placing on top'. But because waves go down as well as up, the effect of combining the motion of two waves is not always a bigger motion; it can be less, too. In adding waves, more can mean less, as well as more.

It's worth pointing out that the simple adding together of waves doesn't invariably work. On a rough day at sea, high wave crests break making 'white horses'. The superposition is not linear: the motions don't simply add to one another. But for light and radio waves the principle works extremely well.

Phase differences

When two waves come together at a place, the result depends on what the two waves are doing. If they are going up and down together, in step with one another, the result is an oscillation with amplitude equal to the sum of the two. But what if one wave is going up as the other goes down, if they are out of step? Then the resulting oscillation is the difference of the two, so can be smaller than either. It even makes a momentary 'flat' if the two are equal in amplitude.

The word **phase** is often used to describe a stage in a change which cycles around, for example 'the phases of the Moon'. The phase of a wave motion says where it is in its wave cycle. Two waves doing the same things at the same moment are said to be **in phase**. They have no **phase difference**. Two waves

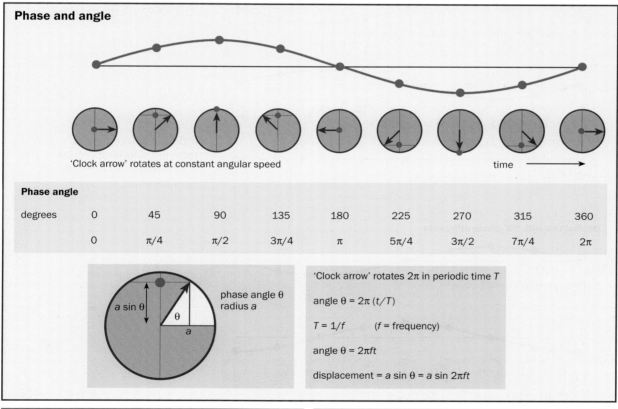

Phase and angle

'Clock arrow' rotates at constant angular speed   time ⟶

Phase angle

degrees	0	45	90	135	180	225	270	315	360
	0	$\pi/4$	$\pi/2$	$3\pi/4$	π	$5\pi/4$	$3\pi/2$	$7\pi/4$	2π

phase angle θ
radius a

$a \sin \theta$

'Clock arrow' rotates 2π in periodic time T

angle $\theta = 2\pi (t/T)$

$T = 1/f$ (f = frequency)

angle $\theta = 2\pi ft$

displacement = $a \sin \theta = a \sin 2\pi ft$

- Colours in thin films
- Overlapping ripples

- Introducing phasors

Superposition and phase difference

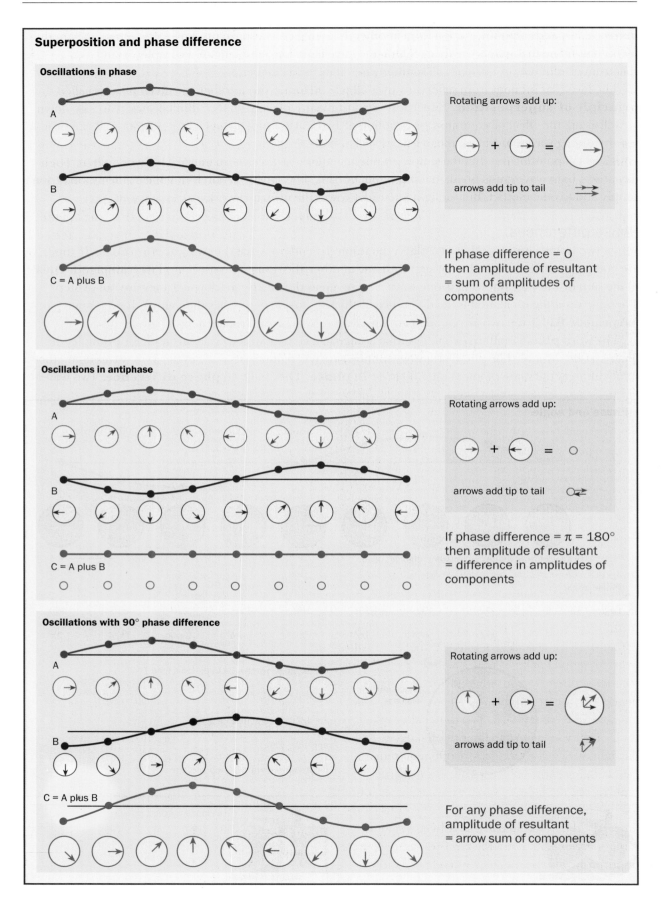

Oscillations in phase

A

B

C = A plus B

Rotating arrows add up:

$\bigodot\!\!\rightarrow$ + $\bigodot\!\!\rightarrow$ = $\bigodot\!\!\rightarrow$

arrows add tip to tail

If phase difference = 0 then amplitude of resultant = sum of amplitudes of components

Oscillations in antiphase

A

B

C = A plus B

Rotating arrows add up:

$\bigodot\!\!\rightarrow$ + $\leftarrow\!\!\bigodot$ = o

arrows add tip to tail

If phase difference = π = 180° then amplitude of resultant = difference in amplitudes of components

Oscillations with 90° phase difference

A

B

C = A plus B

Rotating arrows add up:

$\uparrow$ + $\rightarrow$ =

arrows add tip to tail

For any phase difference, amplitude of resultant = arrow sum of components

doing exactly opposite things at the same moment are said to be **out of phase** or **in antiphase**. These aren't the only possibilities. One wave may be only just a bit behind or ahead of the other, with a small phase difference between them.

Phases or phase differences are simple to measure. They are just angles. A rotating arrow called a **phasor** is used to keep track of where a wave is in its cycle. This arrow 'clock' turns right round once as the wave goes through one whole cycle; both then repeat a new cycle. Between waves which are in phase, the angle is zero. Between waves in antiphase, it is half a complete circle, 180° or π radians.

Oil and soap-film colours

Colours in soap films.

Superposition explains the colours you see in soap bubbles and on oily water. Light falls on the film: some is reflected back from the top surface, some passes into the film and reflects from the lower surface. The light that goes into the film travels further and goes more slowly than the light which comes straight back. There is a delay between light from front and back surfaces. Because of the delay, it can be that peaks and troughs of the two light beams fall on one another. Superposition says that, if so, no light comes back. The film looks dark.

Usually, you view soap films or oil slicks in daylight, which is a spectrum of different frequencies. If the thickness of the film is just right to remove, say, red light, it won't remove other colours and the film looks blue-green. At another place or angle, the difference in paths may be just right to remove green light; there the film looks reddish-blue or purplish. This is why soap and oil films show

secondary colours, and why the colours shift as the film changes thickness and as you move your head.

The same idea, that light which has travelled different distances can be out of step and cancel, is used in the way binary digits are read from a CD-ROM (see chapter 3, page 63).

In all these cases it is the time delay between two parts of the same wave front which matters. For a peak to be delayed so that it falls on a trough of the other part of the wave, it must travel an extra distance equal to half a wavelength of the light in the material of the thin film. Or, of course, any odd number of half-wavelengths. In addition, there may be phase changes at the reflecting surfaces, with peaks coming back as troughs, and vice versa.

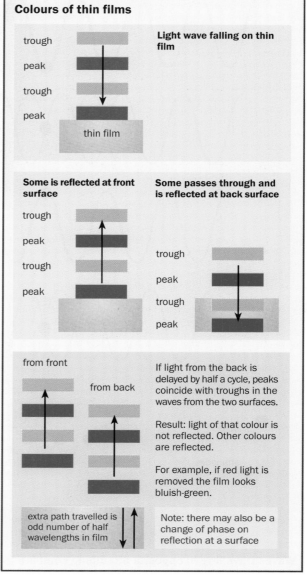

Katherine Blodgett (see chapter 2, page 48) was one of those who found a good use for thin films. The idea was to coat camera and telescope lenses with a suitable fine surface layer, calculated not to reflect light in the middle of the visible spectrum. Such 'bloomed' lenses, often with many layers, look coloured.

More can mean less
Light plus light can give less light. This happens when two beams of light come together but are out of phase. Superposition here leads to 'destructive' interference. The name 'interference' is a bit misleading,

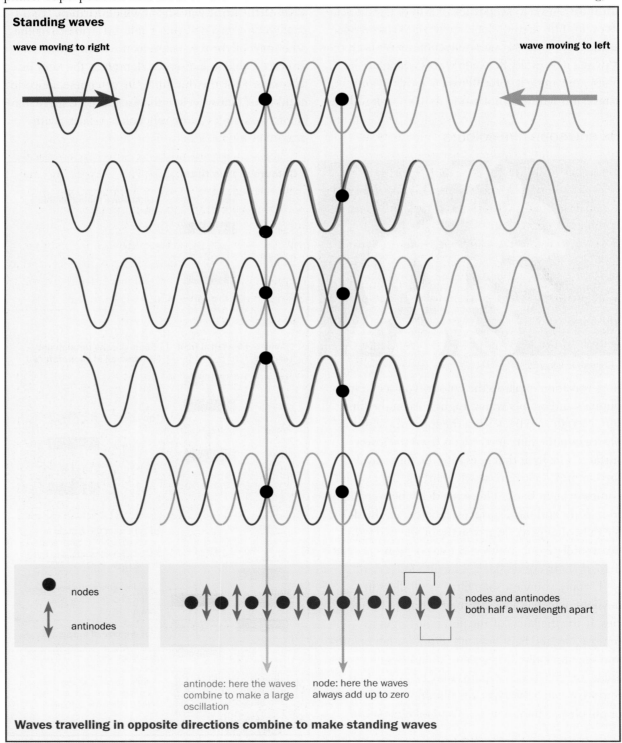

Standing waves

wave moving to right   wave moving to left

● nodes

↕ antinodes

nodes and antinodes
both half a wavelength apart

antinode: here the waves
combine to make a large
oscillation

node: here the waves
always add up to zero

Waves travelling in opposite directions combine to make standing waves

because the whole thing works by the waves not taking the slightest notice of one another, but just adding together to give very little. However, the name is very common and you will meet it frequently. Of course, when waves are in phase, light plus light gives even more light. This is also called interference (often 'constructive' interference).

It is not just light waves which undergo interference as a result of the superposition of two waves. You can observe the same effect in microwaves, radio waves, sound waves and water waves.

You see interference effects when waves come together with a definite phase difference. For this reason, you don't see steady interference patterns on the surface of a choppy sea because the phase differences between the waves are continually changing. Waves only show stable interference effects when they are **coherent**. Light from a laser is coherent over quite long distances, and so easily produces interference.

Using white light, containing a range of wavelengths, you can only see interference patterns when the path differences are small. This is why thick window panes don't show interference colours. For a soap film, the angles at which one colour is removed are very different from those at which a different colour is removed. But for a window pane, places where different colours are removed all overlap.

Standing waves make music

Take a moment and hum the first few notes of a favourite tune. Imagine it being performed. If you are imagining a lead guitar solo you will be thinking of a very different sound to, say, a flute solo; these instruments would sound different even if they were playing the same tune. Not only can the principle of superposition explain why different instruments sound different, it can also explain why instruments can make a tune at all.

The sequence of notes is created on a musical instrument by way of **standing waves**. You may have seen standing waves on a slinky spring or perhaps on a vibrating cord illuminated with a stroboscope. You get standing waves when waves travel in opposite directions through one another. The standing waves on the strings of guitars, and in the pipes of trumpets and clarinets are all examples of superposition: ones which make music.

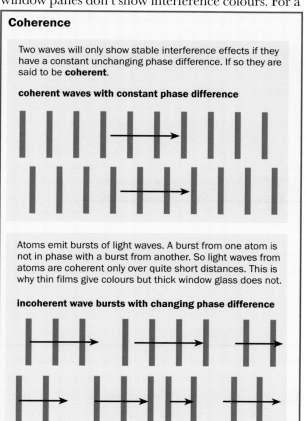

Coherence

Two waves will only show stable interference effects if they have a constant unchanging phase difference. If so they are said to be **coherent**.

coherent waves with constant phase difference

Atoms emit bursts of light waves. A burst from one atom is not in phase with a burst from another. So light waves from atoms are coherent only over quite short distances. This is why thin films give colours but thick window glass does not.

incoherent wave bursts with changing phase difference

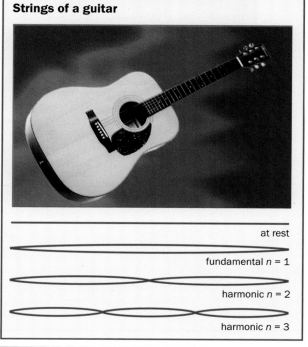

Strings of a guitar

at rest

fundamental $n = 1$

harmonic $n = 2$

harmonic $n = 3$

● Computer animation: Superposition and standing waves

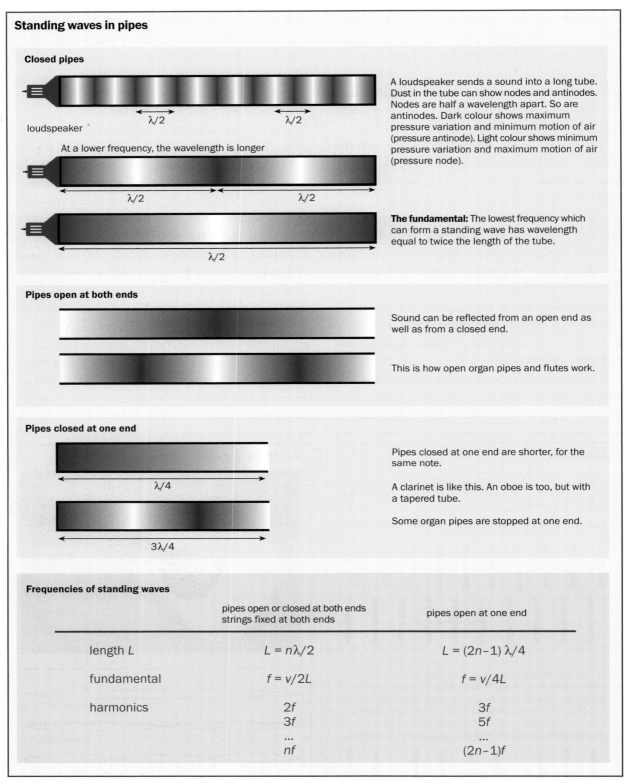

Standing waves in pipes

Closed pipes

loudspeaker

$\lambda/2$ $\lambda/2$

A loudspeaker sends a sound into a long tube. Dust in the tube can show nodes and antinodes. Nodes are half a wavelength apart. So are antinodes. Dark colour shows maximum pressure variation and minimum motion of air (pressure antinode). Light colour shows minimum pressure variation and maximum motion of air (pressure node).

At a lower frequency, the wavelength is longer

$\lambda/2$ $\lambda/2$

$\lambda/2$

The fundamental: The lowest frequency which can form a standing wave has wavelength equal to twice the length of the tube.

Pipes open at both ends

Sound can be reflected from an open end as well as from a closed end.

This is how open organ pipes and flutes work.

Pipes closed at one end

$\lambda/4$

$3\lambda/4$

Pipes closed at one end are shorter, for the same note.

A clarinet is like this. An oboe is too, but with a tapered tube.

Some organ pipes are stopped at one end.

Frequencies of standing waves

	pipes open or closed at both ends strings fixed at both ends	pipes open at one end
length L	$L = n\lambda/2$	$L = (2n-1)\,\lambda/4$
fundamental	$f = v/2L$	$f = v/4L$
harmonics	$2f$ $3f$ ... nf	$3f$ $5f$ ... $(2n-1)f$

When you pluck the string of a guitar, waves travel to each end of the string. They bounce back and forth between the ends of the string. These waves going in opposite directions produce a standing wave on the string, and it vibrates in one or more characteristic patterns. The vibrating string sets the body of the guitar vibrating, and the large surface area of the body efficiently transmits the sound into the air, so that an audience can easily hear it.

A variety of instruments. What is the source of vibrations for each instrument? How is the length of the standing wave changed to produce different notes?

If the player shortens the string by holding it down against a fret, the wavelength of the standing wave is shortened and the note produced goes to a higher frequency.

The note you hear is the frequency of the longest wave that will fit on the string, whose wavelength is twice the length of the string. This is called the **fundamental** frequency. But the quality of the guitar sound is produced by this sound being mixed with higher frequencies from shorter wavelength standing waves called **harmonics** (see chapter 3, page 70).

Standing waves can also be formed in air. They make the notes in wind instruments. When you blow into a recorder the air passing the thin lip ('fipple') below the mouthpiece forms rapidly spinning eddies. This sends a train of pressure waves down the hollow bore of the recorder, which reflect from the open end at the bottom of the instrument. As with the guitar, waves travel up and down the instrument, and make a standing wave inside. The standing wave picks out one particular note, depending on the length of the open bore, usually the fundamental. But if the player blows harder, higher frequencies are present in the eddies around the lip, and a new standing wave of shorter wavelength can be picked out. Uncovering holes along the recorder effectively changes the length of the tube and so changes the wavelength of the standing wave. Once again, if the standing wave shortens, the frequency of the note rises.

The particular quality of sound of an instrument depends on the precise mixture of harmonics in the note, which the player can modify a little. It also depends on other things, such as the time the note lasts and the way it starts and stops.

Other kinds of standing wave include those in rod-like aerials to receive television signals (chapter 3, page 68). People living beside inland lakes have puzzled about how sometimes the water can rhythmically rise and fall—the effect is called a 'seiche' by the French speakers living near the Lac Leman in Switzerland. It is produced by the water in the lake slopping slowly forwards and backwards, in a huge standing wave on the two-dimensional lake surface. You will have seen a seiche yourself if you have ever had a spill when carrying a shallow dish filled with water across to the sink. Standing waves in the cavity of the throat explain the quality of the human voice.

● Acoustics of rooms

Try these

1　A source of sound waves of frequency 570 Hz emits a note of wavelength 0.6 m in air at 20°C. What is the speed of sound at this temperature?

2　A stationary wave is formed in a string with antinodes every 150 mm. What is the wavelength of the standing wave?

3　A loudspeaker points directly at a wall 3 m away and emits a note of frequency 680 Hz. A standing wave is formed. If the speed of sound is 340 m s^{-1} what will be the separation between minimum intensities (nodes)?

4　A steady note of frequency 850 Hz from a works siren is heard by two people standing in an open field. One is 4.0 m nearer the siren than the other. If the speed of sound in air is 340 m s^{-1} what is the path difference in metres between the waves reaching the two workers? How many wavelengths is this?

5　Organ pipes can produce very low notes. About how long would a stopped organ pipe closed at one end have to be to produce a note of 30 Hz? (speed of sound 340 m s^{-1})

Answers 1. 342 m s^{-1} **2.** 300 mm **3.** 0.25 m **4.** path difference 4 m so 10 wavelengths **6.** 2.8 m

YOU HAVE LEARNED

- That when waves superpose their oscillations add.

- That waves can arrive at a point along different paths and can have a phase difference when they meet.

- That phase differences can be measured as angles, using rotating 'clock arrows' (phasors).

- That thin film interference patterns and standing waves can be produced by the superposition of waves.

- That two waves of the same frequency combine by adding their amplitudes if in phase; by subtracting their amplitudes if in antiphase.

- That coherence is necessary for a stable interference pattern.

- That waves on strings or in pipes can form standing waves, and that this explains how the notes produced consist of a fundamental frequency mixed with harmonics.

amplitude, frequency, coherence, phase, phase difference, phasor, interference, superposition

6.2 What is light?

So far in this chapter we have described light as a wave. But as it turns out, the questions 'waves of what?' or 'waves in what?' lead into deep and interesting issues which physicists have had to face, including, surprisingly, questions about the nature of space and time.

Early ideas about light

Go back 2000 years. How would you have explained the iridescent shimmer of a butterfly wing or a beetle's back? Or any colour at all? Or even light itself? Light, though everywhere around us, has always been a mystery. Over the millennia some of the greatest names in philosophy and science, including Aristotle, Galileo, Descartes, Huygens and Newton, and writers including Goethe, have had their say in trying to account for it.

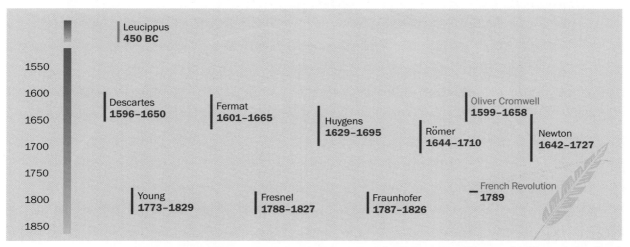

Some of the earliest ideas about light go back to the Greek philosopher Leucippus of Miletus who lived around 450 BC. He speculated that maybe we see objects because their shapes travel across to our eyes. The idea was that very fine layers, like layers of skin off an onion, were continually peeling off and travelling away from things. Philosophers called these imagined layers *eidola*; later, the Romans called them *simulacra*. Needless to say, the notion raised more questions than it answered. What were *eidola* made of, and how did they get into the eye? The idea may seem very strange, but at least Leucippus had it the right way round; many others thought of seeing as if it were like touching, supposing that it is done by 'visual rays' coming from the eyes, and returning to them from objects. They thought that an animal's eyes gleaming in the dark might be a glimpse of such rays. And Leucippus was facing an important mystery not fully resolved even today: how do objects 'out there' in the world somehow 'get into our minds' when we look at them? Or do we deceive ourselves into thinking that they do?

Ahead of his time: Ole Römer and the speed of light

In 1675 the Danish astronomer Ole Römer (1644–1710), working at the Royal Observatory in Paris, became puzzled by the strange motion of the moons of Jupiter. He found that they kept a slowly varying timetable as they went round the planet. Over about six months they would gradually fall further and further behind schedule, and then in the rest of the year catch up again.

There had to be an explanation, and Römer thought of it. He realised that the moons could seem to get behind schedule if light from them took a finite time to reach the Earth from Jupiter. Then because the distance the light had to travel increased day by day as the Earth in its orbit around the Sun got further and further from Jupiter, the moons would seem to show up later and later. When the Earth started to get closer to Jupiter again, the moons would seem to catch up again on their timetable.

In September 1676 Römer told the French Academy of Sciences that a forthcoming eclipse of one of Jupiter's moons would be ten minutes later than calculated. On 9 November the eclipse was observed; exactly ten minutes late. On 22 November 1676 he explained his idea in detail and calculated that light takes about 22 minutes to cross the Earth's orbit, this being the shift in the moons' seeming timetable over one year.

All Römer knew was that the Earth's orbit is '22 minutes worth' of light-time. Not having a good value for the diameter of the Earth's orbit, he couldn't actually estimate the speed of light. Later, with evidence that the Earth's orbit is about 22 000 Earth diameters across, the Dutchman Christiaan Huygens calculated the speed. With a modern figure for the diameter of the Earth this gives a speed of 280 000 km s^{-1}.

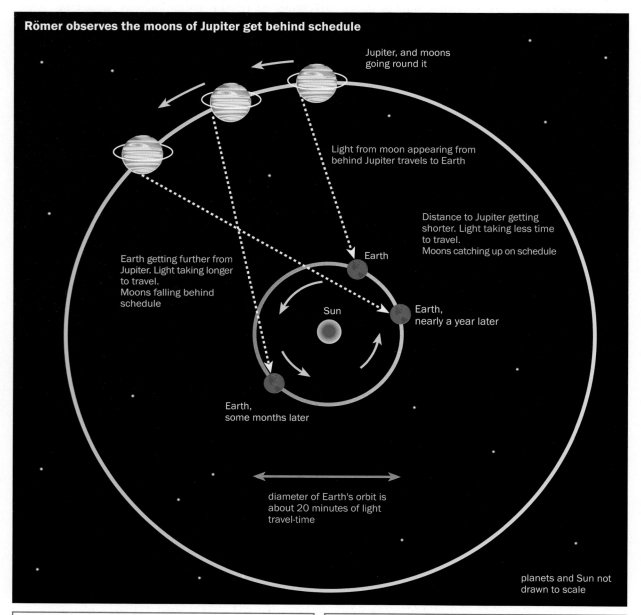

Römer observes the moons of Jupiter get behind schedule

Jupiter, and moons going round it

Light from moon appearing from behind Jupiter travels to Earth

Distance to Jupiter getting shorter. Light taking less time to travel. Moons catching up on schedule

Earth

Earth getting further from Jupiter. Light taking longer to travel. Moons falling behind schedule

Sun

Earth, nearly a year later

Earth, some months later

diameter of Earth's orbit is about 20 minutes of light travel-time

planets and Sun not drawn to scale

● Historical attempts to measure the speed of light

● Measuring the speed of light

Many people shrugged off these results; such a speed was so unimaginably large, they said, that it might as well be infinity. That's what the great philosopher Descartes had said thirty years before, after all.

But Römer's work wasn't so easily dismissed. Twenty odd minutes to travel one of the biggest distances people could then imagine was pretty quick, but definitely not nothing at all as Descartes' opinion required it to be. And the result raised new questions. Why that particular speed? What might light be made of which gives it this speed? As it happens, these questions lead straight into the deepest questions of modern physics. But they also highlight the importance of the deceptively simple question: What is light?

Newton imagines particles

Isaac Newton (1642–1727) began his investigations into the nature of light in 1664 after buying a prism from a trinket stall at a fair. Newton finally published all his work on light in 1704, in his book *Opticks*. The combination of mathematics and experiment found in *Opticks* became a model for future generations of investigators and can be instantly recognised as 'scientific' to this day, even though people no longer believe all its conclusions. *Opticks* is famous for the well-known words,

> *'Are not the Rays of Light very small Bodies emitted from shining Substances?'*

in which Newton gave support to a moving particle picture of light. But the reader of *Opticks* soon finds that Newton actually mixes together particle ideas with something very like wave ideas. He clings to the idea of particles, but suggests that they act by setting up vibrations in matter:

> *'Do not several sorts of Rays make Vibrations of several bignesses, which according to their bigness excite Sensations of several Colours, much after the manner that the Vibrations of the Air, according to their several bignesses excite Sensations of several sounds?'*

(remember that in Newton's time the word 'several' meant 'different'). For 'bigness' read 'wavelength', and you are pretty much in a world of waves.

Newton knew of many wave phenomena, including colours in thin films—in fact he invented a way ('Newton's rings') for measuring them accurately. He asks what might transmit vibrations from the Sun to the Earth as fast as Römer had shown they must travel. When he tries to account for what we now call interference effects, he says that particles of light must have 'alternate Fits of easy Transmission and easy Reflexion', which is close to saying that like waves they have phases. However, on the particle side of the argument, he saw that refraction in glass could only be explained if the particles were pulled towards the glass and speeded up inside it. It was not until 1850, a century and a half later, that Jean Foucault and Armand Fizeau

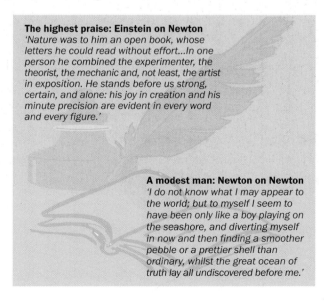

The highest praise: Einstein on Newton
'Nature was to him an open book, whose letters he could read without effort...In one person he combined the experimenter, the theorist, the mechanic and, not least, the artist in exposition. He stands before us strong, certain, and alone: his joy in creation and his minute precision are evident in every word and every figure.'

A modest man: Newton on Newton
'I do not know what I may appear to the world; but to myself I seem to have been only like a boy playing on the seashore, and diverting myself in now and then finding a smoother pebble or a prettier shell than ordinary, whilst the great ocean of truth lay all undiscovered before me.'

both managed to show directly that light actually goes slower in water than in air, not faster (see chapters 2 and 4). Newton had it the wrong way round.

Newton himself did not come down conclusively on one side or the other, but popular explanations such as the curiously titled *Newtonianismo per le dame* (Newtonianism for the ladies), written by Francesco Algarotti in 1735, decided that he had. They told the world that Newton had believed in a travelling particle picture, and for long after his death it was widely assumed to have his backing. In retrospect, 'the dead hand of Newton' may well have hindered the development of alternative pictures of light.

Huygens imagines waves

Christiaan Huygens (1629–1695) was an independent-minded and prolific Dutchman born into a well educated family of diplomats. Not only did he deduce a value for the speed of light from Römer's observations, he also made new observations on Saturn's ring system, discovered Saturn's largest moon Titan, studied the mechanics of pendulums and invented accurate clocks.

It was from Huygens that, two years after Römer's announcement about Jupiter's wayward moons, the French Academy of Sciences again

heard a revolutionary development in thinking. Huygens stood before them to present his ideas on the wave nature of light. Römer was present, and heard his ideas of the finite speed of light used to introduce Huygens' wave concept.

Huygens argued from commonplace observation. He noted the 'obvious' fact that light doesn't get in the way of light; that two torches each illuminate the other as if the other were not shining. He pointed out that travelling light particles would be expected to hit one another, so that the light from one torch should collide with the light from the other. He even remarked that this made it hard to see how two people could look one another in the eye at the same moment. The observation that water waves pass through each other without hindrance—or in modern terms, that they superpose—led Huygens to picture light as a wave motion.

This picture left some difficult questions to answer: 'If light is a wave then what is it a wave *in*?' and 'How can the wave picture explain the fact that light travels in straight lines, and is reflected and refracted?'

Ethereal matter

Huygens thought of light as a longitudinal wave, like sound. He drew pictures of how the wave must travel: very tiny, very hard particles in contact which pass the wave motion from one to the next. Arguing by analogy, Huygens describes how this works with spheres arranged in a line:

"one finds, on striking with a similar sphere against the first of the spheres, that the motion passes as in an instant to the last of them...and one sees that the movement passes with an extreme velocity which is the greater, the greater the hardness of the spheres".

Huygens drew this picture of the 'ether'. He pointed out that ball A and ball D could both hit the row of balls, and that the movement of either, passed through balls B and C, is rapidly transmitted to the other.

> **TRAITE**
> # DE LA LVMIERE.
> Où font expliquées
> *Les caufes de ce qui luy arrive*
> Dans la REFLEXION , & dans la
> REFRACTION.
> *Et particulierement*
> Dans l'etrange REFRACTION
> **DV CRISTAL D'ISLANDE,**
> Par Monfieur CHRISTIAN HUYGENS, Seigneur de Zeelhem,
> *Avec un Difcours de la Caufe*
> **DE LA PESANTEVR.**
>
> *A LEIDE,*
> Chez PIERRE VANDER AA, Marchand Libraire.
> MDCXC.

Title page of Huygens' Treatise.

You can probably see where his argument is going. The velocity of light had just been measured and light went far, far faster than anything else. Therefore, what it goes through (the 'ether') must be composed of particles far, far harder than anything else. This gives him a problem. How do we all move through this mass of hard particles without noticing? How do the planets go through it without slowing down?

The idea of space filled with a material substance to carry light waves has not survived, though it was a long time dying. But another of Huygens' ideas has proved tremendously fruitful. It is his idea of 'Huygens' wavelets' to explain how waves re-create themselves.

Huygens' wavelets

Dip a stick into a pool of water. A ripple spreads out evenly in all directions. Huygens' imagined light spreading in a similar manner:

"[light] spreads, as sound does, by spherical surfaces and waves: for I call them waves from their resemblance to those which are seen to be formed in water when a stone is thrown into it, and which present a successive spreading as circles, though these arise from another cause and are only in a flat surface."

Christiaan Huygens (1629–1695). Huygens, born in Holland to a family of diplomats, did work in France at the Royal court. His *Traite de la Lumière* is addressed in the introduction, to the French king.

Waves, according to Huygens, come from every part (A, B, C) of the candle flame. A little distance from the candle they all seem like spheres centred on the candle.

Huygens had a brilliantly simple idea about how waves get from place to place, changing form as they go. It was that every point on a wave front acts just as a stone dropped in water and sends out circular ripples in every direction. The new wave front is just the places where the wavelets are in phase with one another, and so combine to re-create the wave. Everywhere else, the wavelets from one part of the wave are out of phase with those from another, and the result is nothing.

The wavelet idea is more radical than it looks. It denies that waves simply roll grandly forward just as if they know where to go, even if that's how it looks. It says that you have to think of wavelets from the wave as spreading everywhere, but adding up to nothing everywhere except where the wave actually

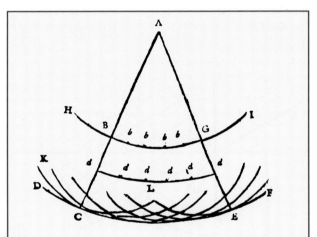

Huygens tries to show how a spherical wave expands. His idea is that the wave front DCEF is made of wavelets from the earlier wave front HBGI, starting from all the places like b,b,b,b on that wave front.

goes. Waves propagate, according to Huygens, because of superposition of wavelets. A similar idea has surfaced again in modern quantum physics (chapter 7), and has further changed ideas about what light is.

Huygens did something else impressive. He proved that, on his wave theory, the bent path of refracted light is the path that takes the shortest time, even though longer in distance than the direct straight path. Thirty years before, the mathematician

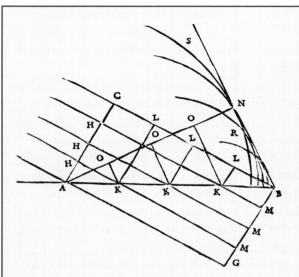

Huygens explains how a wave is reflected. The incoming (plane) wave is CA. From A a wavelet expands, becoming the circle S,N centred on A. Parts of the wave front CA labelled H,H,H arrive later at the mirror, at points K,K,K, and produce wavelets such as R and L. All these wavelets combine to produce the reflected plane wave B,N. The pure geometry ensures that the angles of incidence and reflection are equal.

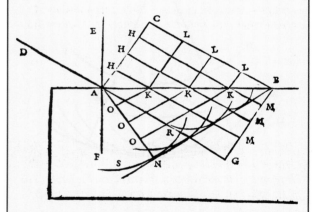

Huygens explains refraction. The incoming wave is CA. Wavelets start out into the glass from points such as K,K,K. They go more slowly in the glass than in air (points O,O,O in glass are closer than points L,L,L in air), so the wave front BN in the glass is swung downwards on the diagram. The geometry leads to Snell's law (chapter 4, page 94).

- Huygens' project
- Ripple tank images

Fermat had had the same idea, but lacked a theory of light to account for it.

You may have been struck by how the ideas of Newton, Huygens and others depend on simple analogies with well-known things. Newton thought that light was *like* travelling particles, whereas Huygens believed it was *like* a travelling wave of sound. You might ask yourself whether this is 'scientific'. If you think not, then where else can people start except with what they know?

What did Huygens achieve?

Huygens' imagined wavelets made it possible to see how a wave propagates, how light meeting a mirror obeys the law of reflection:

angle of incidence = angle of reflection

and how light passing from air into glass or water obeys Snell's law of refraction:

$$\frac{\sin i}{\sin r} = \text{refractive index}$$

He also made a prediction: that the speed of light should be smaller in water or glass than in air. Nobody could test this at the time, but in the end it turned out to be right. The speed measured by Fizeau and Foucault (pages 141–142) did change in the right way and by the right amount. All explained just by superposing wavelets.

But Huygens' idea also posed a huge problem. What could these waves be travelling in? Huygens called it 'ether', but this just gave the problem a name, not a solution. What could the 'ether' be like? How could the stars and planets slip unresistingly through such a substance?

Sir George Stokes (1819–1903) suggested that the ether was like a thick mixture of glue and water that acts like a solid for rapid vibrations but allows slow movement with relative ease.

You can make your own 'Stokesian ether' by adding a small amount of water to a few table-spoons of custard powder or cornflour to make a very thick cream. This 'non-Newtonian fluid' will allow you to move a spoon through it slowly but will resist quicker movement. If you get the mixture just right you can shatter it with a hammer! Soon, however, the electromagnetic theory put paid to such mechanical models of the ether (see page 151).

Try these

1 Römer interpreted his findings to suggest that light takes about twenty minutes to cross the diameter of the Earth's orbit. If the Earth is 1.5×10^8 km from the Sun, what does this give for the velocity of light?

2 Why, knowing that light going from air to glass is refracted towards the normal, would Newton have supposed that particles of light must travel faster in glass than in air?

3 Suggest one argument in favour of picturing light as a stream of particles, and one against.

4 Suggest one argument in favour of picturing light as a wave, and one against.

5 Draw a sketch to show how Huygens would have explained the refraction of light coming *out of* glass *into* air.

6 The first measurement of the speed of light done on the Earth's surface, by Fizeau in 1849, timed flashes of light out and back over the 5.36 mile (8.6 km) distance from Montmartre to Suresnes. What time interval did Fizeau have to be able to measure?

Answers 1. 250 000 km s^{-1} **6.** 60 μs approximately

YOU HAVE LEARNED

- The finite speed of light could be detected in delays in the regular motion of moons round Jupiter.

- During the 18th century, a moving particle model of light was widely held, and its supporters claimed the authority of Newton.

- Huygens' wavelet idea explains wave propagation as the superposition of wavelets from every point on a wave front, so creating a new wave front.

- Huygens' wavelets explain reflection and refraction of light, again using superposition of wavelets.

- Wave theories of light such as Huygens' had the problem of trying to describe the 'ether', in which the waves were supposed to propagate.

speed of light, wave speeds, Huygens' wavelets, reflection, refraction

6.3 Wave behaviour understood in detail

Thomas Young revives wave theory

In 1801, at the Royal Society in London, a medical doctor gave an important physics lecture. Today you would expect a physicist to give the talk and the doctor to stick to medicine. But two centuries ago, people really admired men and women who knew a bit (often a great deal) about many subjects. The doctor on the podium on that day, Thomas Young (1773–1829), besides medicine knew a dozen languages, wrote poems, did physics experiments, and helped to decipher the Egyptian hieroglyphics on the Rosetta Stone. But by all accounts he wasn't at all good at explaining himself clearly, which made it hard for him to get his ideas respected.

The subject of his lecture was the wave theory of light. In England, most people, partly out of respect for Newton, held to the particle theory. Young wanted to show that the wave theory was alive and kicking. He reminded his audience of Newton's many observations of what look like wave phenomena—colours in soap films, bright and dark fringes.

Young went on giving such lectures. One of his most convincing demonstrations was very simple: the 'two-slit' experiment, which along with the modulus of elasticity (chapter 4) now bears his name. Young shone light through two barely separated pinholes onto a distant screen. On the screen appeared a regular pattern of bright and dark fringes. Off-centre, the light from one pinhole had further to go than the light from the other, and the two could get out of step. Where they did, the screen would be dark. A little further off-centre the two would be in step again, and the screen would be bright. Particularly convincingly, if one pinhole was covered up the bright and dark fringes vanished.

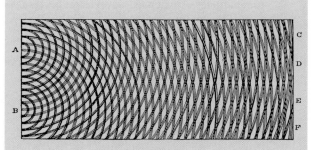

Thomas Young's own drawing to explain the two-slit interference pattern (from *Lectures on Natural Philosophy* 1807). Look along the page from the left and you can see curved lines where peaks and troughs coincide, their ends marked C,D,E,F.

Looking at the fringes, you are in a way looking at the tiny light wavelength 'blown up' in scale. The spacing x between the fringes is the wavelength λ multiplied by the ratio L/d, which is at least 1000 and possibly 10 000, since the slit–screen distance L is over a metre and the slit spacing d is less than a millimetre. That is:

$$x = \lambda \frac{L}{d}$$

This is how Young expressed the idea of wave superposition:

"When two Undulations, from different origins, coincide either perfectly or very nearly in Direction, their joint effect is a Combination of the Motions belonging to each".

To understand Young in modern terms, for 'Undulations' read 'waves', for 'Combination of the Motions' read 'superposition'.

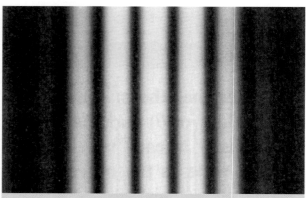

The fringe pattern from two slits.

Portrait of Thomas Young (1773–1829). Young was a polymath—linguist, archaeologist, poet and medical doctor. He lectured badly, it seems, and scornful reviews of his presentation of the wave theory of light are to be found in journals of the time.

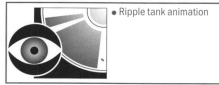

● Ripple tank animation

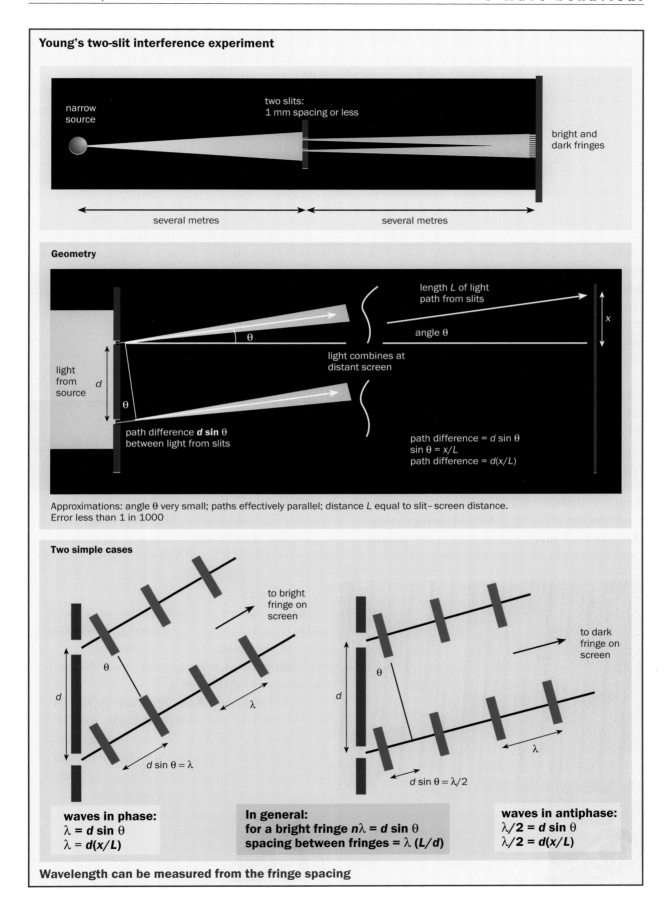

Young's two-slit interference experiment

narrow source

two slits:
1 mm spacing or less

bright and dark fringes

several metres

several metres

Geometry

length L of light path from slits

angle θ

x

θ

light combines at distant screen

light from source

d

θ

path difference $d \sin \theta$ between light from slits

path difference = $d \sin \theta$
$\sin \theta = x/L$
path difference = $d(x/L)$

Approximations: angle θ very small; paths effectively parallel; distance L equal to slit–screen distance. Error less than 1 in 1000

Two simple cases

to bright fringe on screen

to dark fringe on screen

θ

d

λ

$d \sin \theta = \lambda$

θ

d

λ

$d \sin \theta = \lambda/2$

waves in phase:
$\lambda = d \sin \theta$
$\lambda = d(x/L)$

In general:
for a bright fringe $n\lambda = d \sin \theta$
spacing between fringes = $\lambda (L/d)$

waves in antiphase:
$\lambda/2 = d \sin \theta$
$\lambda/2 = d(x/L)$

Wavelength can be measured from the fringe spacing

Young's audiences were sometimes far from impressed. A reviewer wrote savagely of:

"...the feeble lucubrations of this author, in which we have searched without success for some traces of learning, acuteness, and ingenuity, that might compensate his evident defiency in the powers of solid thinking, calm and patient investigation, and successful development of the laws of nature, by steady and modest observation of her operations."

It was a youthful French engineer Auguste Fresnel (1788–1827) who really changed scientific opinion, through brilliance and hard work. His prize-winning memoir of 1819 reports very accurate experiments, explaining each experiment in detail from the wave point of view, as exactly as possible. The mathematical ideas he developed are still used today (and his invention, the Fresnel lens, sits in every classroom overhead projector).

Gratings and spectra

We began this chapter with the blue wing of the Morpho butterfly (page 129). The many tiny platelets on the wing, arranged in regular rows, sharply reflect just one colour. It is the regular rows that do the trick, and the same idea was used by Joseph Fraunhofer in 1821 when he became the first to obtain spectra using a **diffraction grating**. Thousands of fine lines ruled very close together on glass or on a mirror replaced the rows of platelets.

Micrograph of the ruled surface of a grating.

The spiral track on a CD-ROM produces the same effect, seen in the rainbow colours of light reflected from it.

With the equivalent of thousands of slits, light waves scattered from each line can—at just the right angle—all be in phase with one another for one particular wavelength. This is the wavelength of the colour you see. As a new scientific instrument the diffraction grating proved immensely useful. Indeed it made it possible for the first time to know what the stars are made of: the elements they contain show up with their spectral fingerprints in the light from the star. Famously, the element helium was discovered in the spectrum of the Sun before being produced in the laboratory.

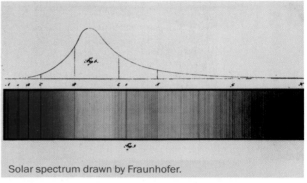
Solar spectrum drawn by Fraunhofer.

The grating is now an essential tool of modern physics, pure and applied. High precision gratings carefully and accurately ruled on metal or glass are expensive to make, but cheap multiple plastic copies made from a single original are good enough for many purposes.

Spectra taken with gratings need not be in the visible region—spectra of infrared and ultraviolet light can be obtained too. These often require reflection gratings, if the grating would absorb the radiation passing through it.

At the beginning of the 20th century, regular rows of atoms in materials started to be used as gratings for the much smaller wavelength x-rays, first to show that x-rays have wavelike behaviour, but later as a tool to investigate the arrangements of the atoms themselves inside materials.

● Diffraction and interference for pleasure

● Using a CD as a reflection grating

Diffraction grating

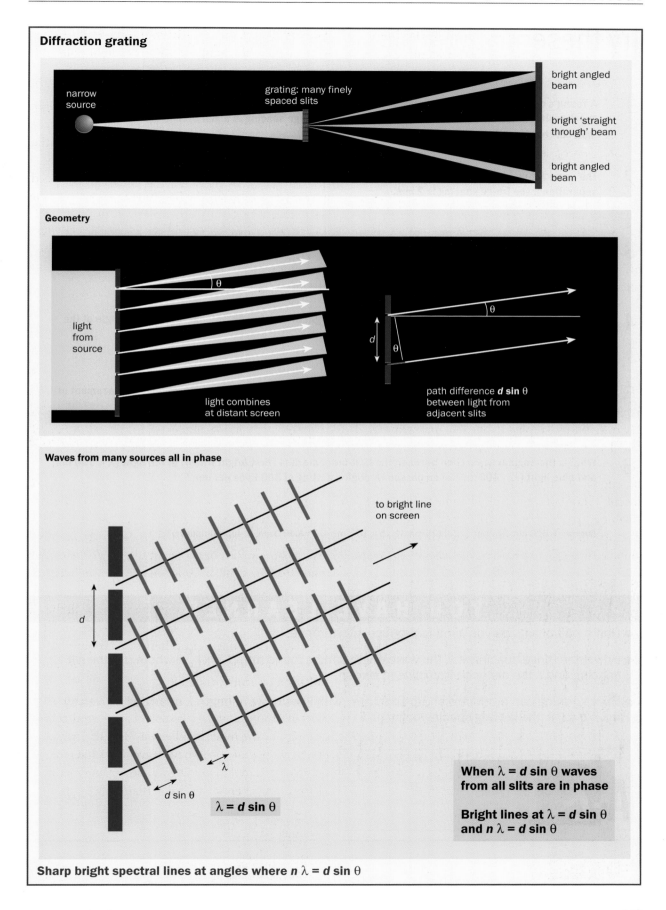

narrow source

grating: many finely spaced slits

bright angled beam

bright 'straight through' beam

bright angled beam

Geometry

light from source

θ

light combines at distant screen

θ

d

θ

path difference $d \sin \theta$ between light from adjacent slits

Waves from many sources all in phase

to bright line on screen

d

λ

$d \sin \theta$

$$\lambda = d \sin \theta$$

When $\lambda = d \sin \theta$ waves from all slits are in phase

Bright lines at $\lambda = d \sin \theta$ and $n \lambda = d \sin \theta$

Sharp bright spectral lines at angles where $n \lambda = d \sin \theta$

Try these

1 A Young's double slit experiment is set up using a semiconductor laser. The screen is placed 3 m away from the two slits, which have a separation of 0.4 mm. If the fringe spacing is 5 mm what is the wavelength of the light emitted from the laser?

2 The same set-up is used but the slits are replaced with a pair of unknown separation. Calculate the slit separation if the fringe spacing is 2 mm.

3 What is the slit spacing of a grating marked '80 lines per mm' ?

4 A grating with spacing 1×10^{-5} m is illuminated with light of wavelength 550 nm. What is the angle of the first-order fringe (i.e. when $n = 1$)?

5 The same grating is illuminated with light of an unknown wavelength that gives an angular displacement of 10° between the first and last of three bright fringes. What is the wavelength?

6 What is the angular separation between the first-order maxima (first bright fringe) of red light ($\lambda = 700$ nm) and blue light ($\lambda = 400$ nm) when passed through a grating of 300 lines per mm?

Answers 1. 670 nm **2.** 1mm **3.** 0.0125 mm (1.25×10^{-5} m) **4.** 3° **5.** 870 nm **6.** approximately 5°

YOU HAVE LEARNED

- How a pair of slits can be used to produce interference.

- How if the fringe spacing is x, the wavelength can be found from $\lambda = x \left(\dfrac{d}{L} \right)$ where d is the slit spacing and L the distance from slits to screen.

- How a grating can produce a sharp spectrum, with lines of wavelength λ at angles θ given by $n \lambda = d \sin \theta$, the grating spacing being d.

double slit interference, diffraction, gratings

6.4 Looking forward

The wave picture of light went from strength to strength, except for the awkward fact that nobody knew what light was waves *of*. The next big step was taken in 1865 by James Clerk Maxwell, who followed up a speculative idea put forward twenty years before by Michael Faraday. Faraday had written in an essay *Thoughts on ray vibrations*,

> *The view which I am so bold as to put forth considers, therefore, radiation as a high (frequency) species of vibration in the lines of force...It endeavours to dismiss the ether, but not the vibration.*

By 'lines of force' Faraday means a field, like that which you will have seen around a magnet. No more material 'ether'; no imaginary wave-transmitting substance undetectably filling empty space. Just vibration. Maxwell turned this visionary idea into what is now called the electromagnetic theory of light. And as you know, modern communications have sprung from that idea (chapter 3).

The next chapter (chapter 7) reveals that the story is not yet over. Quantum physics takes a hint from Faraday, who wanted to throw away the ether and keep just the vibration, but goes further still. The idea is to strip away all the stories about waves in this or in that, and keep just 'how to work things out'. The rotating arrows called phasors, introduced at the start of this chapter, are just the tool that is needed.

Diffraction into a harbour.

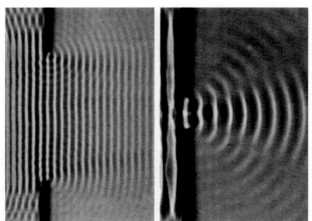
Diffraction of water ripples through apertures.

Waves spread out from apertures

Both Young and Newton knew that the Italian Francesco Grimaldi (1618–1663) had long before shown that if you put an obstacle in front of a small source of light, then the light bends into the shadow, with coloured fringes around the shadow of the edges. This is diffraction.

Fresnel put Huygens' wavelet idea to careful mathematical use, and gave a precise explanation of the patterns of light and dark, in convincing quantitative detail. This makes it possible to predict, for example, by how much the beam of a microwave transmitter from a dish aerial will spread.

Huygens' wavelet idea says, 'imagine every part of the aperture (radar dish, or slit through which light passes) as sending out wavelets. Then for a given place receiving the waves, just add up all the wavelets'. The phasor picture makes this recipe easy to follow. Every wavelet has a phasor, which rotates at the wave frequency as it travels, going once around for every cycle and so once around for every

● Computer animation: Superposition and standing waves

● Beats: Continually varying phase difference

Diffraction at a single aperture

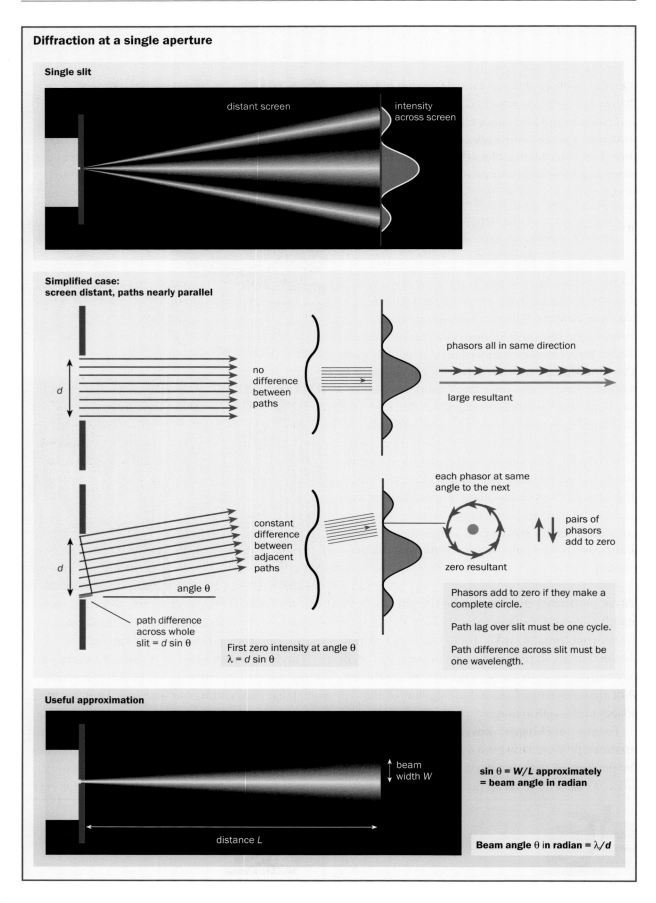

Single slit

distant screen

intensity across screen

Simplified case: screen distant, paths nearly parallel

d

no difference between paths

phasors all in same direction

large resultant

d

constant difference between adjacent paths

each phasor at same angle to the next

pairs of phasors add to zero

zero resultant

angle θ

path difference across whole slit = $d \sin \theta$

First zero intensity at angle θ
$\lambda = d \sin \theta$

Phasors add to zero if they make a complete circle.

Path lag over slit must be one cycle.

Path difference across slit must be one wavelength.

Useful approximation

beam width W

distance L

$\sin \theta = W/L$ approximately = beam angle in radian

Beam angle θ in radian = λ/d

wavelength of path travelled. Then you just add up the phasors at each place to get the total wave amplitude at that place.

Going to a place straight ahead of the aperture and a long way away, all the wavelets have travelled equally far, and so have the same phase. Their phasor arrows point in the same direction, and they add up to a large amplitude.

Going to a place off to one side, wavelets from one edge of the aperture have further to go than ones from the other side. So a phase lag builds up between wavelets coming from successive parts of the aperture. The diagram of phasors added together starts to curl up. Just suppose it curls into a complete circle. The last phasor bites the tail of the first. Then the wavelets all add up to nothing at that place. For this to happen the total phase lag across the whole aperture must be one cycle, or in terms of path difference, one wavelength. The result is that the edge of the beam where the waves first have zero intensity is at angle θ where

$$\lambda = d \sin \theta \text{ or, if the angle is small, where beam angle } \theta \text{ in radian} = \frac{\lambda}{d}$$

Your eye has just such an aperture: the pupil. It is interesting to think how this affects the resolution of your eye, and whether it is the size of the pupil or the micrometre spacing of rods and cones in the retina which actually most limits the resolution.

Resolution of instruments

You may have wondered why radio telescopes and radar signalling equipment are built with such large reflecting dishes. Whether the signal is being sent out or is being received through the aperture of a dish, the angle of the beam in radians is of the order of magnitude λ/d. (For a circular aperture in place of a narrow slit, a more exact value is $1.22 \, \lambda/d$.) This means that a radio telescope or radar aerial can't tell apart distant objects which are closer than this angular distance apart. The objects cannot be **resolved**. Modern radio astronomy solves the problem by using different telescopes as far apart as possible on the Earth, but linked to act as if they were a single virtual telescope. Diffraction similarly limits the resolution of optical telescopes and microscopes.

Stripped down thinking

The phasor picture gives a rather simple description of diffraction; indeed Fresnel used much the same reasoning in analysing a whole range of diffraction and interference phenomena. The diagram (opposite) explaining diffraction at an aperture is peculiar in a very significant way. It makes almost no mention of waves: just spinning phasor arrows. Superposition just becomes adding arrows tip to tail. Even the 'wavelength λ' is just the distance over which a phasor arrow turns round once.

So what, you may ask? Surely the phasor arrows are designed to work as they do because light is a wave motion? In the next chapter, there is a surprise in store. We will, in describing the world of quantum physics, turn this argument back to front. Light, we shall say, exchanges energy in lumps called photons. Why then does light behave like a wave? The answer will be that light seems to behave as a wave because phasor arrows governing the photons behave as they do.

As sometimes happens in physics, what started as an abstract mathematical device for calculating (spinning phasor arrows) turns out to give a deeper view of reality, by unhooking the mind from one picture (waves) and opening up new ways of seeing things.

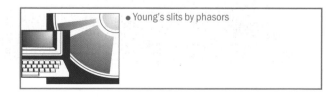
• Young's slits by phasors

Try these

1   Ocean tides are sinusoidal waves with a period of approximately 12 hours. If the difference between high water and low water is 2 m, use a phasor diagram drawn to scale to predict (a) the water height above the mean sea level 1.5 hours before high tide, (b) how far the water rises in the two hours following low tide, (c) the mean rate of rise or fall of the water level between 1 hour before and 1 hour after half tide.

2   A 1 m diameter radar dish on a telecommunications tower transmits radar waves of wavelength 10 mm. Estimate the beam width at a range of 10 km.

3   Sketch the phasor arrows for bright and dark fringes in a Young's two-slit experiment.

4   The pupil of your eye has a diameter of about 5 mm. The wavelength of light is of the order of 500 nm. What is the limit on the angular resolution of your eye set by the size of the pupil? To what width on the retina, at a distance of say 25 mm, does this correspond? How does this compare with the spacing of rods and cones in the retina?

5   The 300 m diameter radio telescope dish at Arecibo in Puerto Rico is sunk in a shaped hollow in the ground. Suppose that unevenness in its surface makes 100 mm be the shortest wavelength which is usable. Estimate the angular resolution which can be expected at this wavelength.

6   The amplitude of waves through a single narrow slit, in the straight through direction where the phasors line up in a straight line, is A. What fraction of A is the amplitude at an angle to the side where the phasors curl up into a semicircle so that the resultant is the diameter of the circle?

7   Look at the ripples diffracted by the narrower slit in the photograph on page 151. The angle between the direction of the first minimum and the straight through direction is about 30°. Estimate the ratio of the width of the slit to the wavelength.

Answers 1. (a) 0.7 m above mean level, (b) 0.5 m, (c) 0.5 m per hour. **2.** 100 m; **3.** alternate pairs of arrows, pointing in the same and in opposite directions **4.** 10^{-4} radian, 2.5 μm, roughly comparable **5.** 3×10^{-4} radian **6.** $(2/\pi)A$ **7.** about 2

YOU HAVE LEARNED

- How waves passing through an aperture of width d spread out, typically over an angle θ given by $\lambda = d \sin \theta$

beats, diffraction, single slit, phasor

- That these and other wave effects can be analysed in a simple way by using the phasor picture. Phasor arrows have length equal to the wave amplitude, and rotate at the wave frequency.

- How wave superposition is represented by adding phasor arrows tip-to-tail.

Summary checkup

✓ Superposition

- Interference, diffraction and standing waves can be produced by the superposition of waves.

- Two waves of the same frequency combine by adding their amplitudes if in phase, or by subtracting their amplitudes if in antiphase.

- Coherence is necessary for a stable interference pattern.

- Standing waves on strings or in pipes have a fundamental frequency, with harmonics which are whole number multiples of the fundamental.

- Wavelength can be calculated from Young's two-slit experiment using $\lambda = x\,(d/L)$ where x is the fringe spacing, d is the slit separation and L is the distance from the slit to the screen.

- Wavelength can be calculated for diffraction gratings using the relationship $n\lambda = d\sin\theta$ where θ is the angular displacement of a bright fringe, n is the fringe order and d the slit separation.

- Waves passing through an aperture of width d spread out, typically over an angle θ given by $\lambda = d\sin\theta$

✓ The nature of light

- The finite speed of light could be detected in delays in the regular motion of Jupiter's moons.

- During the 18th century, a moving particle model of light was widely held, and its supporters claimed the authority of Newton.

- Huygens' wavelet idea explains wave propagation as the superposition of wavelets from every point on a wave front, so creating a new wave front. It explains reflection and refraction of light.

- Wave theories of light such as Huygens' had the problem of trying to describe the 'ether', in which the waves were supposed to propagate.

- Young and Fresnel provided support for the wave theory by calculating the shape of interference and diffraction fringes in a range of experiments.

- Fraunhofer and others made precision observations of spectra using the diffraction grating as a tool.

✓ The phasor picture

- Phase differences can be measured as angles, using rotating 'clock arrows' (phasors).

- Phasor arrows have length equal to the wave amplitude, and rotate at the wave frequency.

- Wave superposition is represented by adding phasor arrows tip-to-tail.

- The phasor picture enables simple quantitative calculations to be done starting from Huygens' wavelet idea.

Questions

Speed of light = $3 \times 10^8 \, m \, s^{-1}$

1 **The back of a certain beetle looks bright blue from a certain angle, but not from other angles. Assume that this is due to diffraction by ridged layers of plates on the surface of the beetle's back, laid out like overlapping tiles.**

(a) Make a sketch showing how blue light might be scattered from successive ridges so as to be seen strongly reflected at a particular angle.

(b) What is the order of magnitude of the spacing between ridges?

(c) If the beetle were to evolve so as to look red at the same angle, how would the layout of the ridged 'tiles' on its back have to change?

2 **A two-slit interference experiment is to be set up for an Open Day. It will be done with laser light, wavelength 600 nm. The closest the slits for the laser experiment can be ruled is 0.3 mm.**

(a) How many wavelengths apart are the slits?

(b) If the screen is set 3 m from the slits, will visitors be able to see the bright fringes distinctly? What will be their spacing?

(c) A visitor calculates that the ratio of the fringe spacing to the distance from slits to screen is 1 to 500. Explain how this is connected to the slit spacing, measured in wavelengths.

(d) The experiment takes up a lot of space. If the distance from slits to screen were reduced to 1 m, what effect would that have on the fringe spacing?

(e) It is suggested that a second two-slit experiment should be set up alongside, using microwaves of wavelength 30 mm, as a scaled-up version of the laser experiment. A student objects that the two slits would need to be ridiculously far apart. Is the objection reasonable?

(f) After some discussion, the slits for the microwave version are put 150 mm apart. Make a scale drawing to show the direction in which to expect the first 'bright fringe'.

(g) Very roughly, what would be the fringe spacing in the microwave experiment if the detector was set 1 m from the slits?

3 **The waves carrying a microwave communications link have a frequency of 15 GHz. They are made into a beam by a reflecting dish 2 m in diameter.**

(a) At what rate does the phasor representing the wave rotate?

(b) How far does the wave travel in one rotation of its phasor?

(c) How is the answer to (b) related to the wavelength of the waves?

(d) Draw a sketch using phasors to show how, although the beam is strong along the axis of the dish, it can fall to near zero at a small angle to the axis. (Treat the 2 m dish as a slit 2 m wide.)

(e) Estimate the width of the beam from the dish at a distance of 1 km.

4 **Standing waves, interference, diffraction, colours in thin films, are all examples of wave superposition. Wave superposition is therefore involved in music and art as well as in scientific and technological work. Choose one example of an application of wave superposition effects in any field, and use it to explain:**

(a) what the superposition effect is in your example, and why you count it as a case of superposition.

(b) the use to which superposition is put in your example.

(c) how to calculate any quantities relevant to your example.

(d) reasons for choosing the frequencies or wavelengths of the waves in your example.

5 **Imagine that you have heard a lecture by Römer on his ideas about light. The audience is invited to raise any objections they have.**

(a) Remind the audience of one thing Römer is likely to have said, and state your objection, with a reason.

(b) Suggest how a member of the audience might respond to a lecture given by either Huygens or Newton.

6 **The reed of a clarinet effectively closes it at one end, the other end of the pipe being open.**

(a) Sketch the standing wave in the clarinet when it is playing the lowest note possible.

(b) Sketch the standing waves for the first two harmonics above the fundamental.

(c) Relate the length of a given type of clarinet to the range of notes it can play. Start from knowledge of, or an estimate of, either length or frequency, and relate it to the other.

7 Quantum behaviour

Fundamental particles of matter—electrons, quarks, neutrinos and others—act like nothing you have seen before. But they all share the same remarkable style of acting: 'quantum behaviour'. Photons, the fundamental particles of light, do it too. In this chapter we will:

- explain why light needs to be thought of as photons
- use photons as a simple example of quantum behaviour
- show that electrons behave in that way too
- prepare the way for the modern picture of fundamental particles

7.1 Quantum behaviour

Quantum behaviour is important and fundamental. It accounts for the way atoms behave, and for the ways in which they combine to form molecules. It underlies how the most fundamental particles behave. Essential at these very small scales, quantum behaviour shows up on the large scale too: for example, the superconducting coils in magnets used in medical scanning.

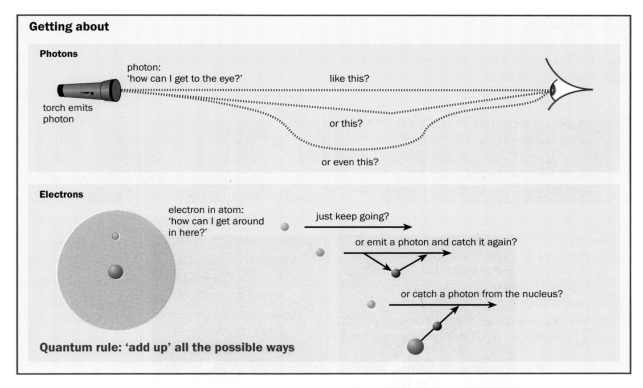

'...Dick Feynman told me about his ... version of quantum mechanics. "The electron does anything it likes," he said. "It just goes in any direction at any speed, forward or backward in time, however it likes, and then you add up ...." I said to him, "You're crazy." But he wasn't.'

Freeman Dyson in H Woolf (ed.), *Some Strangeness in Proportion* (1980) Addison Wesley, p 376

Quantum things have their own way of behaving, completely unlike the behaviour of everyday objects. The rule for quantum behaviour is very simple. Its simple anarchic command is, 'Try everything possible; explore all paths'. Then all the possibilities are added up in a special way, and used to predict events. The details can get complicated, even fiendish, but the conception is simple. Simple, grand and very general.

New ideas always seem peculiar at first, but you get used to them. That's how it is with quantum behaviour. Start young and you stop worrying sooner!

We will limit most of the story of quantum behaviour to a very basic problem—the problem of getting from A to B, of how particles travel from one place to another. But the same basic ideas work for how particles change into one another, or interact with one another. In other words, they work for all of quantum physics.

We will tell first the story of the quantum behaviour of photons, because it's the simplest story. Later we will show how electrons behave in essentially the same way, though it's a bit more complicated because of their mass and their charge.

Light arrives randomly in lumps

Take a photograph in dimmer and dimmer light. You might expect that less light will just give a fainter photograph. But it doesn't. Instead, the photograph breaks up into randomly arranged grains, just as if the light energy were arriving randomly in lumps. Turn up the brightness, and the lumps of light arrive thick and fast, and produce the smooth-looking picture you expect. The lumps, or **quanta**, of light are called **photons**.

These images give a good mental picture of how photons arrive, and what is meant by saying that their arrival is 'random'. It means that where the final picture is bright, there is a high probability of arrival of photons. Where it is dim, the probability is low. But these are probabilities, not certainties. A photon can always appear where the probability is low or fail to appear where it is high. But such events, although they do happen—look at the images—are rare. Overall, more photons arrive where the probability is high and fewer arrive where it is low. As we shall explain, the rule of quantum behaviour, 'Try everything possible', predicts these probabilities.

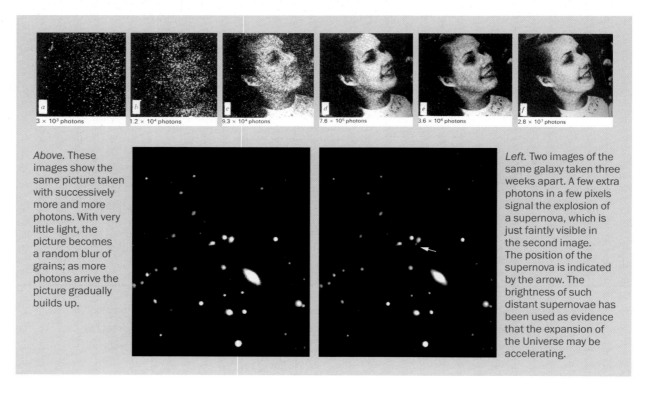

3×10^3 photons 1.2×10^4 photons 9.3×10^4 photons 7.6×10^5 photons 3.6×10^6 photons 2.8×10^7 photons

Above. These images show the same picture taken with successively more and more photons. With very little light, the picture becomes a random blur of grains; as more photons arrive the picture gradually builds up.

Left. Two images of the same galaxy taken three weeks apart. A few extra photons in a few pixels signal the explosion of a supernova, which is just faintly visible in the second image. The position of the supernova is indicated by the arrow. The brightness of such distant supernovae has been used as evidence that the expansion of the Universe may be accelerating.

A real example comes from astronomy. One night in 1997, a few hundred extra photons from a very distant galaxy arrived in a telescope in the Atacama desert in Chile. They registered on only a few pixels in the image of the galaxy made by its ccd camera, but enough to indicate the flaring-up of a powerful supernova. Those few hundred photons were just large enough in number to be distinguished from a random fluctuation in the other photons coming in from the galaxy.

Your eyes are not quite sensitive enough to detect the lumpiness in light: it takes half a dozen photons to trigger a rod on the retina. They can be detected by a photomultiplier, in which the arrival of a single photon is amplified to become a measurable current pulse of electrons.

The photomultiplier

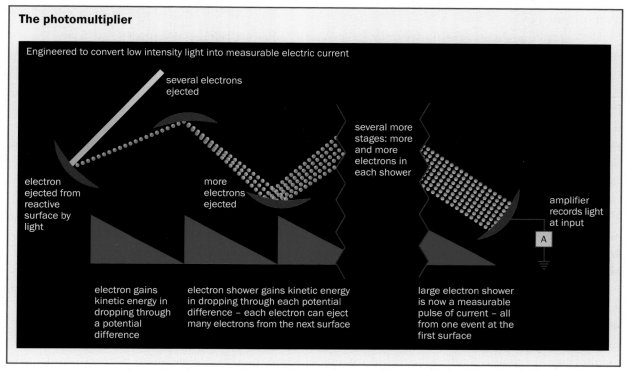

Engineered to convert low intensity light into measurable electric current

several electrons ejected

electron ejected from reactive surface by light

more electrons ejected

several more stages: more and more electrons in each shower

amplifier records light at input

A

electron gains kinetic energy in dropping through a potential difference

electron shower gains kinetic energy in dropping through each potential difference – each electron can eject many electrons from the next surface

large electron shower is now a measurable pulse of current – all from one event at the first surface

Gamma ray photons arrive in more energetic quanta than those of visible light. A quite simple detector lets gamma rays from a radioactive source be heard arriving as distinct clicks. As the source is brought closer to the detector, the clicks merge into a continuous noise. Listening to gamma rays arrive, you hear:

'click..........click....click..........click...............click..click........click'

They come at random, sometimes several close together, sometimes with longer gaps. The arrival of one quantum is completely independent of the arrival of the one before or after—no pre-determined railway timetable for them. This pattern is found everywhere in quantum behaviour, like the emission and detection of alpha and beta radiation from radioactive sources. Like photons, they too come at random. Beneath events seeming to happen smoothly lie events happening one by one at random with a certain probability. Seeming smoothness is down to averaging. It is a bit like going out in the rain. You can't tell when or where the next raindrop will fall, but if you stay out you will get wet. Strong sources close up—like heavy rain—give a rapid average rate of arrival, while for weak ones far away—like a scattered shower— photons come less often. The randomness extends to the whole of quantum physics. Atoms emit photons at random. Nuclei decay at random. But always underneath is a fixed probability which can be calculated.

Images with increasing exposure

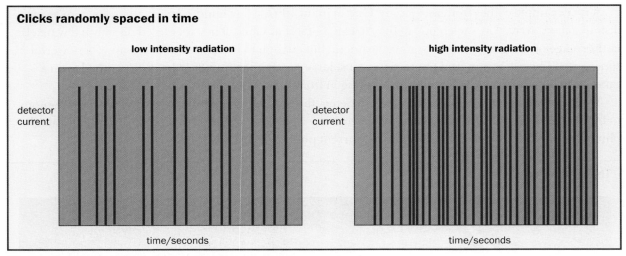

Frequency and quantum energy

It was Max Planck (1858–1947) who first wrote down the connection between the quantum of energy of electromagnetic radiation and its frequency. The relationship is very simple:

Electromagnetic radiation of frequency f is emitted and absorbed in quanta of energy E where

$$E = hf$$

in which h is the Planck constant, $6.6 \times 10^{-34}\,\text{Js}$

The Planck constant h relates the quantum energy to the frequency of the radiation. Higher frequency radiation comes in quanta of larger energy. The quantum energy of x-rays or gamma rays is big enough to knock an electron out of an atom, which is why they can damage living tissue. The quantum energy of radio waves is tiny, which is why at most they stir electrons in metal aerial rods into gentle motion.

When Planck stumbled on this idea he was trying to 'fix' the awkward fact that the wave theory of light gave the absurd result that even slightly warm objects should emit x-rays. But he was never really happy with it. A century later, however, quantum ideas are essential to a physicist's everyday work. Atoms and molecules emit or absorb photons of characteristic energy and frequency. Their sharp spectra are used in: monitoring car exhausts; finding the rate of rotation of stars; analysing the pigments in a mediaeval painting; identifying substances found at the scene of a crime; mapping a patient's internal organs in a body scanner. The photons finger what is going on.

Steal the wave calculations; forget about the waves

Photons do something else besides arriving and departing in discrete quanta of energy. When there are alternative paths to the same point interference effects appear. Chapter 6 is full of examples of superposition: thin films, Young's slits, gratings, diffraction. Superposition must be fundamental to quantum behaviour.

Here we will take one of the simplest possible illustrations of quantum superposition: Young's two-slit experiment. There are just two paths, going by one slit and going by the other. The rule of quantum behaviour says to the photon, *'Do both!'*. Explore both paths.

This sounds peculiar if the story of photon energy arriving in lumps has led you to imagine a photon as a little parcel travelling like a particle. Such a parcel can't be imagined being in two places at once. So a photon, which explores both paths at once, can't be like a little localised parcel. Lumpiness in energy doesn't mean lumpiness in space. A photon can't simply be a particle.

Waves have no problem in being spread out in space: they send wavelets through both slits, which superpose at the screen. But waves don't deliver energy in well-defined lumps. So a photon can't simply be a wave. Just because photons go everywhere possible doesn't mean that their lumpy energy is spread thinner and thinner.

Evidence for the graininess of light

Light-emitting diodes

LEDs are engineered to drop each electron by a fixed p.d. and to emit a photon of a definite colour. A range of LEDs, of different colours, illustrate the relationship between energy and frequency for photons.

Increase the p.d. until the LED *just* glows. This is the striking p.d.

energy transferred to each electron $= e \times \Delta V$

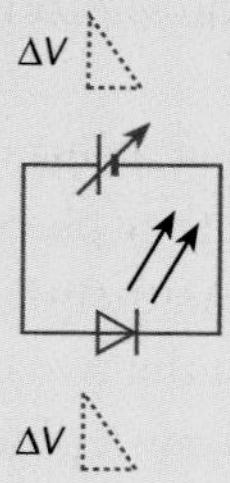

ΔV

energy transferred to each photon $= e \times \Delta V$

ΔV

True as long as the LED does not warm up.

Striking p.d. fixes energy
$E = qV$

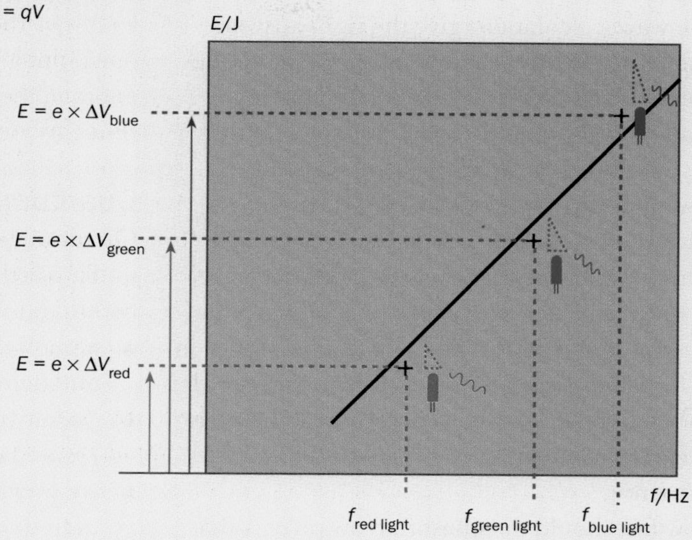

E/J

$E = e \times \Delta V_{blue}$

$E = e \times \Delta V_{green}$

$E = e \times \Delta V_{red}$

f/Hz

$f_{red\ light}$ $f_{green\ light}$ $f_{blue\ light}$

Constant slope, E/f. The number of joules per hertz is uniform for all electromagnetic radiation.
h, the gradient, is 6.634×10^{-34} J Hz^{-1}. More often written as $h = 6.634 \times 10^{-34}$ J s

Photoelectric effect

The ejection of electrons from metals by photons was important in establishing the photon description.

The energy from a single photon is transferred to a single electron.

potential difference just stops electrons

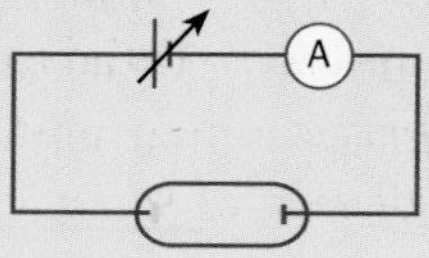

energy to just climb potential hill $= e \times \Delta V$

ΔV

energy transferred by each photon $= e \times \Delta V + \phi$

Stopping p.d. measures electron energy $E = qV$

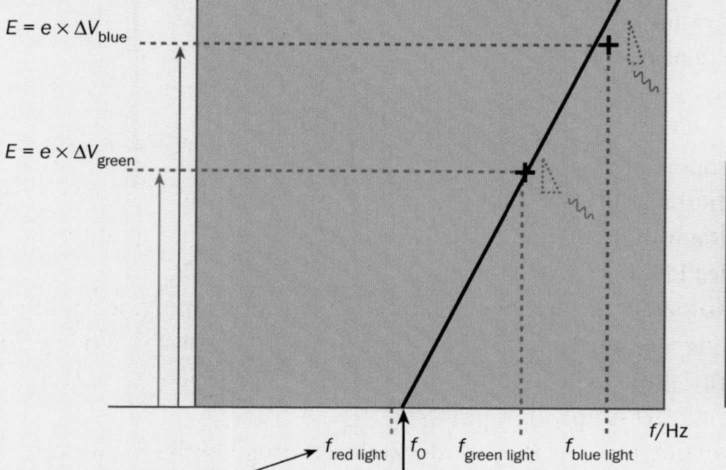

E/J

$E = e \times \Delta V_{blue}$

$E = e \times \Delta V_{green}$

Constant slope, E/f. The number of joules per hertz is uniform for all electromagnetic radiation.

h, the gradient, is 6.634×10^{-34} J Hz^{-1}
More often written as $h = 6.634 \times 10^{-34}$ J s

E/J

metal that gives up electrons more easily

original metal

f/Hz

$f_{red\ light}$ f_0 $f_{green\ light}$ $f_{blue\ light}$ f/Hz

too low a frequency to provide the energy to eject electron

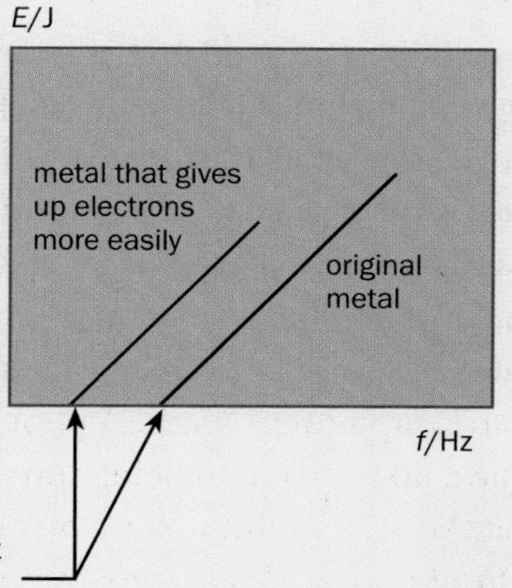

$\phi = hf_0$ energy needed to eject one electron from the metal

Quantum behaviour isn't particle behaviour and it isn't wave behaviour. Quantum objects like photons explore all possibilities, and yet still arrive in lumps.

The wave calculations give the right answers for where there are bright and dark fringes on the screen. So try a bold idea: steal the method of calculation from the wave story, but don't let that make you think that the photons *are* waves. Just borrow what you need, nothing more.

What then is actually *used* in the Young's slits calculation? It is surprisingly little. There are two paths, one from each slit. There is a difference in the lengths of the paths, which gives a difference in phase for light coming along each path and decides whether the screen will be light or dark. The phase at the end of each path can be kept track of by a phasor arrow.

How fast should the phasor arrow turn? What frequency *f* should it be given? Easy: adapt the relationship to photon energy *E* found by Planck:

$$f = \frac{E}{h}$$

That's all. No waves as such. Only path difference and phase difference. Just two spinning phasor arrows that end up pointing in the same or different directions, and must be combined tip to tail into a single resultant.

The resultant phasor decides how bright or dark the screen will be at that place. But in quantum behaviour, 'bright' or 'dark' just means that it is more, or less, probable that a photon will randomly land there. So the length of the resultant phasor arrow is used to calculate the *probability* that a lumpy quantum of energy will arrive.

The probability must end up being proportional to how bright the light is, to its intensity. In the wave calculation, the length of the arrow represents not the intensity of the wave but its amplitude. The intensity—the rate at which energy arrives—is proportional to the square of the amplitude, which is why big waves at sea are so dangerous (for the

reasons, see *Advancing Physics* A2 Student's book, chapter 10). So the quantum probability is calculated by once again stealing from the wave calculation: get the probability by squaring the length of the arrow. Thus if the resultant phasor at one point is three times longer than the phasor at another, the probability of arrival of photons is nine times greater at the first point than at the second.

Quantum behaviour for photons

Quantum behaviour is a new kind of behaviour unlike anything else. For photons, these are its commandments:

- a photon is emitted at a certain place and time, and detected at another place and time
- 'in between', imagine the photon taking *every possible path* between the two events
- for every path, see how a phasor arrow, rotating at frequency $f = E/h$, ends up for the photon going along that path
- add up the phasor arrows, tip to tail, for all possible paths, to get their resultant phasor
- from the square of the resultant phasor, calculate the probability of arrival of a photon.

The phasor sum describes how the different possibilities all combine together, to give a *probability*. That's all there is to it. Not wave behaviour, not particle behaviour, but quantum behaviour.

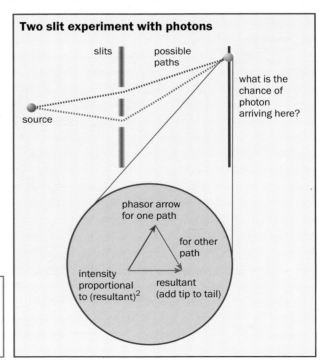

Two slit experiment with photons

slits / possible paths

source

what is the chance of photon arriving here?

phasor arrow for one path

for other path

intensity proportional to (resultant)^2

resultant (add tip to tail)

● A photon explores two holes

How to do quantum calculations

	Place the source and detector. Fix any other obstructions to exploration.	Placing these three things restricts the paths that you can choose.

$$f = \frac{E}{h}$$

f is the frequency of the photon
h is Planck's constant

Fix the frequency at which the phasor will rotate as the photon explores the paths.

f decides the colour of the light, or, more generally, the part of the electromagnetic spectrum involved.

Define a suitable collection of paths for each photon to explore.

The paths which are possible decide the phenomenon you are modelling.

Explore a path by spinning a phasor as it moves along the path at the speed of light.
Freeze the phasor on arrival at the detector to get a final arrow.

This arrow represents one possibility.

Repeat for all remaining paths, starting with a fresh phasor each time.
Record the final arrow for each path.

Each arrow represents one distinct possibility.

Place the arrows nose to tail. The sum of these arrows is the amplitude.

All the possibilities are combined.

Resultant for first position of detector

Square the amplitude to find the chance that a photon ends up at this detector.

The combined possibilities are converted to a probability.

Resultant for next position of detector

Choose another position for the detector and define a new set of paths to do another calculation. Comparing this number with the last predicts the ratio of the photons arriving at each.

Possibilities have combined to give probabilities.
The higher the probability, the more photons arrive.
The more photons that arrive, the higher the intensity.

Try these

Planck constant $h = 6.6 \times 10^{-34}$ J s. In these questions use the rough value 6×10^{-34} J s. Speed of light $= 3 \times 10^8$ m s^{-1}.

1   **What is the energy transferred when a photon of each of the following frequencies is emitted or absorbed? (a) $f = 3$ GHz, (b) $f = 6 \times 10^{14}$ Hz, (c) $f = 1.2 \times 10^{18}$ Hz.**

2   **How many visible photons are produced per second from each of the following lamps? (a) 2 W, (b) 40 W, (c) 100 W. Take the power given to be the power of the lamp in a range of wavelengths near 5×10^{-7} m.**

3   **What is the rate of rotation of the phasor for a photon of each of the following energies? (a) 3×10^{-19} J, (b) 6×10^{-19} J, (c) 12×10^{-19} J.**

4   **A photon of frequency 6×10^{14} Hz explores a path 6 m long. Find the time of travel and the number of rotations of the phasor.**

5   **What is the photon frequency if the phasor makes one complete turn in a path of length 300 mm? What does this tell you about the possibility of bench-scale interference experiments with such photons?**

6   **In a Young's two-slit experiment, the paths from the slits to a place on the screen differ by 900 nm. How many extra turns does the photon phasor make on the longer path, if its frequency corresponds to a wavelength of 600 nm. Will the screen be bright or dark at that place?**

Answers 1. (a) 1.8×10^{-24} J, (b) 3.6×10^{-19} J, (c) 7.2×10^{-16} J **2.** (a) 6×10^{18}, (b) 1×10^{20}, (c) 3×10^{20} **3.** (a) 5×10^{14} Hz, (b) 1×10^{15} Hz, (c) 2×10^{15} Hz **4.** 2×10^{-8} s; 1.2×10^7 **5.** 1 GHz; wavelength is 300 mm—interference experiments easily arranged, as in chapter 6 **6.** one and a half turns; dark

YOU HAVE LEARNED

- That quantum behaviour is unique; neither wave nor particle behaviour.

- That photons show quantum behaviour.

photon, quantisation, phasor, quantum amplitude, Planck constant, probability

A-Z

- How to calculate the energy $E = hf$ delivered by a photon.

- That quantum behaviour combines phasors from all possible paths.

- That the probability of an event is got from the square of the resultant phasor amplitude.

7.2 'Many paths' at work

'Explore all paths' sounds like a recipe for anarchy, not for order. But it does have a definite rule: add up the phasor arrows for all the paths. And this rule gives new insights into many of the simple things you already know about light.

Looking glass world

Look in a mirror. Photons come off the mirror surface at the same angle as the angle at which they reach it. No quantum anarchy here, it seems. Not quite so. The 'explore all paths' rule actually explains why that special angled path is favoured.

Think of a source of light and a detector above a mirror. Take a photon from source to detector by any path that hits the mirror. Yes, *any* path, including the ones like those hitting the mirror near the ends, which you know 'don't happen'. On each path, record the trip time in rotations of a phasor, turning at frequency

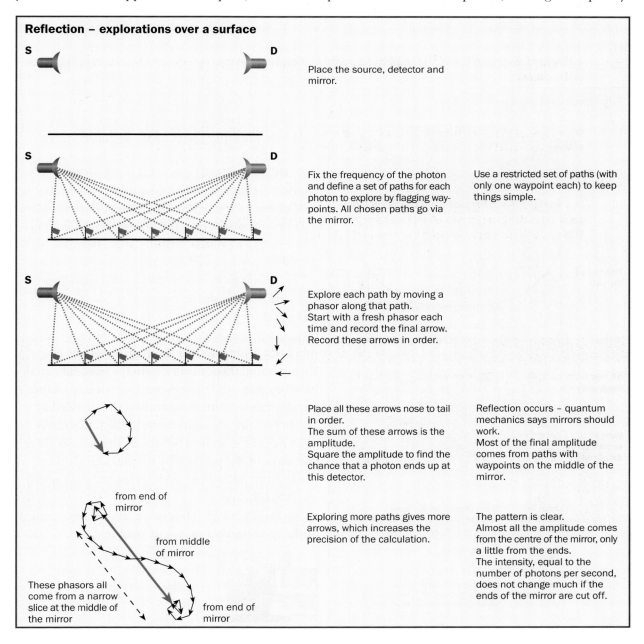

Reflection – explorations over a surface

Place the source, detector and mirror.

Fix the frequency of the photon and define a set of paths for each photon to explore by flagging way-points. All chosen paths go via the mirror.

Use a restricted set of paths (with only one waypoint each) to keep things simple.

Explore each path by moving a phasor along that path.
Start with a fresh phasor each time and record the final arrow.
Record these arrows in order.

Place all these arrows nose to tail in order.
The sum of these arrows is the amplitude.
Square the amplitude to find the chance that a photon ends up at this detector.

Reflection occurs – quantum mechanics says mirrors should work.
Most of the final amplitude comes from paths with waypoints on the middle of the mirror.

Exploring more paths gives more arrows, which increases the precision of the calculation.

The pattern is clear.
Almost all the amplitude comes from the centre of the mirror, only a little from the ends.
The intensity, equal to the number of photons per second, does not change much if the ends of the mirror are cut off.

from end of mirror

from middle of mirror

These phasors all come from a narrow slice at the middle of the mirror

from end of mirror

$f = E/h$, with speed c for the photon. A handy model of this is a 'trundle wheel' that you roll along each path.

You will notice that there *is* something special about the paths near the one where the angle of reflection is equal to the angle of incidence. Phasors for these paths differ little in phase at the detector, and so they more or less line up. But phasors going by other 'wrong' paths differ a lot in phase from nearby ones, so when placed tip to tail tend to curl up. The 'lined up' phasors give a big resultant; the 'curled up' phasors contribute little to the resultant.

This gives a new story about light, which turns what you have learned before on its head. You probably learned the 'law of reflection'; quantum behaviour provides an explanation of that law. The rule of quantum behaviour sends photons every-where, not just along one special path. But the special 'equal angled' path obeying the law of reflection is in fact picked out by the quantum rule for combining phasors from every possible path. This special path and ones very close to it are the only ones for which the phasors line up rather than curl up. Most of the probability of photons arriving at the detector comes from paths fitting closely to the law of reflection. The rules of quantum behaviour predict the law of reflection.

Least time; flat valley bottoms

The paths near where the trip times are equal and the phasors line up are also the *shortest* paths; the ones with the smallest trip time. A graph of trip time has a minimum—it forms a valley. And a valley is flat at the bottom. That is, if the trip time is a minimum, changing the path a little hardly changes the time. Result: the phasors from these paths line up. Quantum behaviour explains reflection and Fermat's least-time principle (pages 143–144), all in one go.

Refraction: going faster, going slower

Chapter 6 shows how Huygens explained refraction, as due to waves slowing down in water or glass. The photon story keeps the slowing down, but once again says to a photon, 'try all paths'. Because of the difference in speed, trip time is now important.

The photons do not sit down beforehand and work out what to do. As usual, they just try everything. But on the paths near the one of shortest time, the phase differences between paths are all small and so the phasors for these paths line up. Paths around those further from the least-time path have big phase differences between them, and the sets of phasors from these curl up when added in. So once again, the anarchic-seeming but actually strict rules of quantum behaviour predict a definite law: most of the probability for photons arriving comes from paths close to the refracted path which obeys Snell's law.

So Snell's law of refraction is also a consequence of the quantum behaviour of photons.

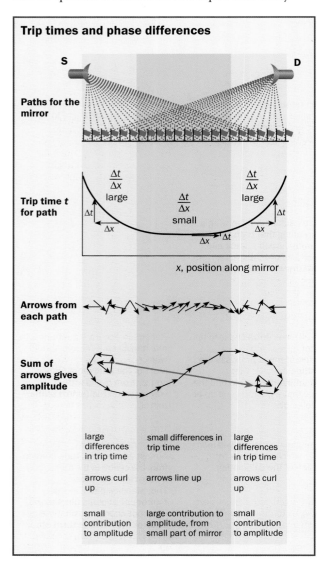

- How exploring paths leads to arrows
- Calculating for a mirror on the screen
- Least time and refraction

Lifeguard – which is the quickest path to get to the swimmer?

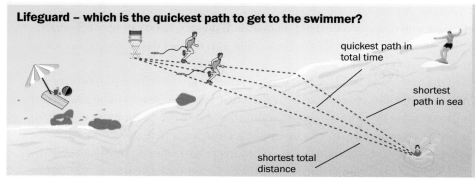

quickest path in total time

shortest path in sea

shortest total distance

A lifeguard is trying to get to a bather in trouble, in the shortest possible time. The lifeguard can run quickly on the beach but can only swim slowly in the sea. The best path is not the direct path. The direct path travels further in the sea than is necessary. The best path is bent, going further on the beach and taking a shorter line in the sea.

Refraction – explorations through a surface

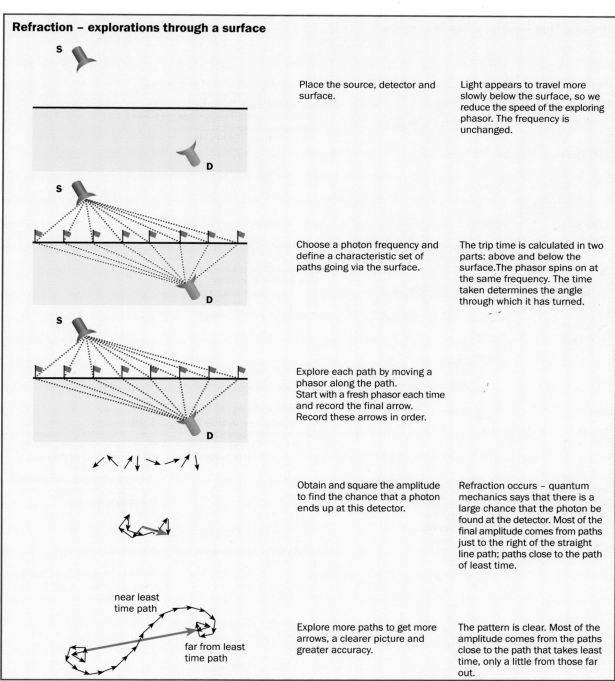

Place the source, detector and surface.

Light appears to travel more slowly below the surface, so we reduce the speed of the exploring phasor. The frequency is unchanged.

Choose a photon frequency and define a characteristic set of paths going via the surface.

The trip time is calculated in two parts: above and below the surface. The phasor spins on at the same frequency. The time taken determines the angle through which it has turned.

Explore each path by moving a phasor along the path.
Start with a fresh phasor each time and record the final arrow.
Record these arrows in order.

Obtain and square the amplitude to find the chance that a photon ends up at this detector.

Refraction occurs – quantum mechanics says that there is a large chance that the photon be found at the detector. Most of the final amplitude comes from paths just to the right of the straight line path; paths close to the path of least time.

near least time path

far from least time path

Explore more paths to get more arrows, a clearer picture and greater accuracy.

The pattern is clear. Most of the amplitude comes from the paths close to the path that takes least time, only a little from those far out.

Seeing straight

It's a remarkable fact that reflection and refraction turn out, in quantum physics, to be examples of superposition, that is, of 'interference'. It is even more remarkable to realise that so is the fact that you can't see round corners.

You may well have guessed the story by now: that light goes in straight lines because photons try every path, including paths which are not straight. You would be right. Paths near the straight line path differ hardly at all in phase. Their phasors line up. The resultant amplitude, and so the probability of arrival, from them is large. Paths far from a straight line differ a lot in phase from their neighbours, so when they are added in the sets of phasors curl up and add little extra. As before, this is connected with the fact that the straight line path is the quickest, so trip times change very little for other paths near that minimum.

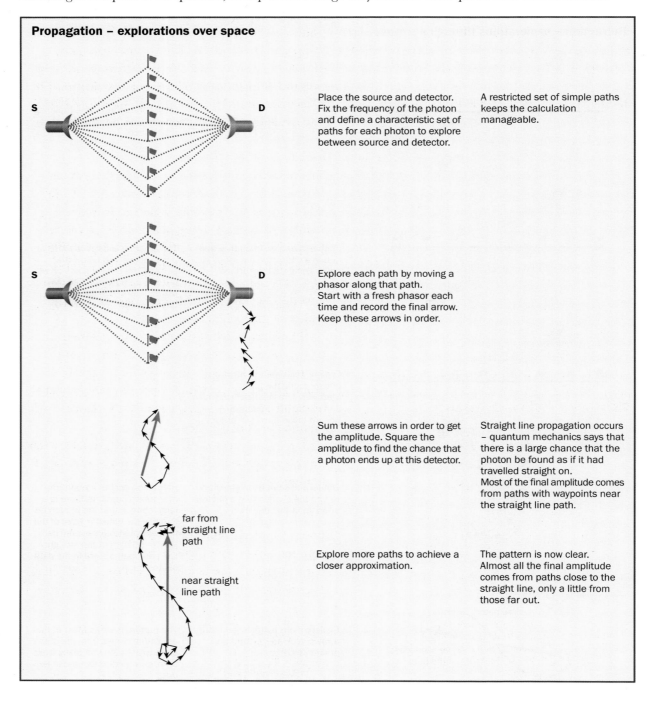

Propagation – explorations over space

Place the source and detector. Fix the frequency of the photon and define a characteristic set of paths for each photon to explore between source and detector.

A restricted set of simple paths keeps the calculation manageable.

Explore each path by moving a phasor along that path. Start with a fresh phasor each time and record the final arrow. Keep these arrows in order.

Sum these arrows in order to get the amplitude. Square the amplitude to find the chance that a photon ends up at this detector.

Straight line propagation occurs – quantum mechanics says that there is a large chance that the photon be found as if it had travelled straight on. Most of the final amplitude comes from paths with waypoints near the straight line path.

far from straight line path

near straight line path

Explore more paths to achieve a closer approximation.

The pattern is now clear. Almost all the final amplitude comes from paths close to the straight line, only a little from those far out.

Engineering with photon paths

A telescope designer, or the optical engineer of a CD player, wants to get a lot of photons from a source to come together at the same place. The photons don't care: they go everywhere. All the designer has to do is to arrange that the trip times along as many different paths as possible are the same. Then the phase differences for those paths are small, their phasors line up, the amplitude and probability are big and—'look, no hands'—the photons arrive in large numbers. All done by superposition.

Easy? Well, sometimes. Take a flat mirror reflecting light back at its source. The trip times via the edges of the mirror are longer than those via the middle. The answer is to bend the flat mirror in a bit at the edges, shortening the paths more and more as they get towards the edges. Bend the mirror just the right amount and the paths can all be the same length. That's why telescopes use curved mirrors.

How about spectacle lenses or a burning glass? The optical engineer must *stop* light travelling in straight lines, and get it to bend. The paths far from the straight one take too long. They can't be made shorter, but instead the straight-through paths can be made *slower*, to get the same effect. How? By putting more glass in the way of straight paths than of up-and-down paths. The solution you know already—a convex lens. All paths through the lens take the same time to go from object to image. The longer ones travel less far through slow-speed glass, and the light arrives in phase from them all.

Getting more from less

The motto for superposition is, 'more can mean less'. Equally true is, 'less can mean more'. Here is how *getting rid of* possible paths can *increase* the number of photons going in a certain direction. Joseph Fraunhofer did it when he made the first diffraction grating (chapter 6, page 148). It sent light of a particular wavelength off in a direction it would not have gone before, a new direction in which all the path phasors line up.

Think about the story for light travelling in straight lines. The off-centre paths curl up and contribute little. That's because they alternately point one way and then the opposite way. So just block off alternate paths, leaving only phasors pointing mainly one way.

If you look at chapter 6, page 149, you'll see that this is another way of telling the story about how a grating works.

Really going everywhere

Look at a small bright lamp through the crack between two of your fingers, and gently squeeze the fingers together. Just before they close, the light blurs out sideways. Angles where there was no chance to

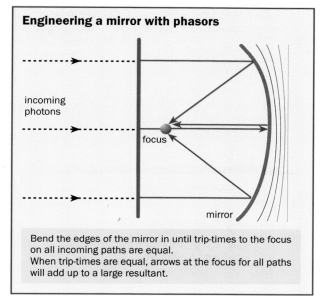

Engineering a mirror with phasors

incoming photons

focus

mirror

Bend the edges of the mirror in until trip-times to the focus on all incoming paths are equal.
When trip-times are equal, arrows at the focus for all paths will add up to a large resultant.

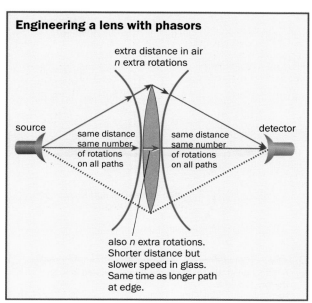

Engineering a lens with phasors

extra distance in air
n extra rotations

source

same distance
same number
of rotations
on all paths

same distance
same number
of rotations
on all paths

detector

also *n* extra rotations.
Shorter distance but
slower speed in glass.
Same time as longer path
at edge.

- Engineering a focusing mirror
- Engineering a lens
- Photons explore a narrow hole

see a photon suddenly get some chance. This is called diffraction (chapter 6, page 152). It happens because photons really do arrive with a probability which takes account of everywhere they might have gone.

Imagine a slit so narrow that all paths from any part of the slit to anywhere else are pretty much the same length, counted in phasor turns. For that, the slit must be narrower than the distance for one phasor turn—one wavelength. Then photon paths at all angles after the slit will have much the same probability for a photon to arrive along one path as along any other. The light will spread out into a uniform blur.

Open up the slit and more paths become possible. Some directions have phasor resultants which are small or zero. The beam behind the slit gets narrower. Open the slit completely and you are back to light travelling in straight lines, because all the other paths add up to zero amplitude.

The result is that the way to make a narrow beam of photons is the opposite of what you might think. Narrow the beam down and it spreads. Open it up and the photons go straight. Astronomers and satellite engineers know it well: to get good directionality make telescopes and aerials as *wide* as possible. That's what makes astronomy expensive.

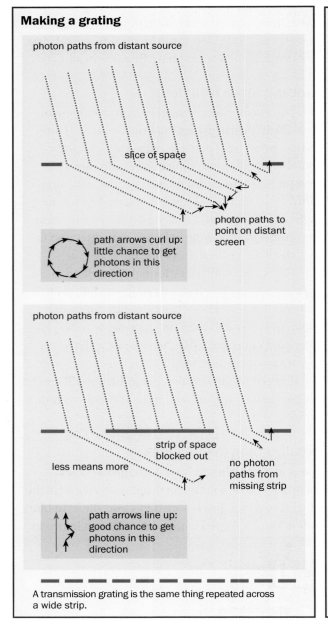

Making a grating

photon paths from distant source

slice of space

photon paths to point on distant screen

path arrows curl up: little chance to get photons in this direction

photon paths from distant source

strip of space blocked out

less means more

no photon paths from missing strip

path arrows line up: good chance to get photons in this direction

A transmission grating is the same thing repeated across a wide strip.

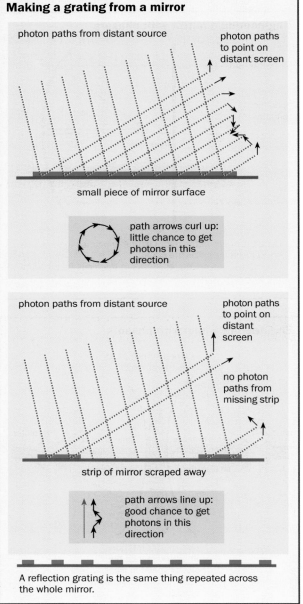

Making a grating from a mirror

photon paths from distant source

photon paths to point on distant screen

small piece of mirror surface

path arrows curl up: little chance to get photons in this direction

photon paths from distant source

photon paths to point on distant screen

no photon paths from missing strip

strip of mirror scraped away

path arrows line up: good chance to get photons in this direction

A reflection grating is the same thing repeated across the whole mirror.

Try these

1   Three photon paths in air from source to detector have the same length. Here is the phasor at the end of one of the paths: →. Draw the phasors for the other two paths.

2   The phasor at the end of one path is →. Draw phasors at the ends of paths on which the phasor (a) makes one extra turn and (b) makes two and a half extra turns.

3   Given fixed source and detector, paths via a plane mirror near the path for which angle of reflection is equal to angle of incidence vary little in length if the angles are slightly altered. Why do such paths contribute substantially to the resultant amplitude for arrival of photons at the detector?

4   The straight line path from source to detector is the shortest path. What effect does this fact have on the phasors at the ends of paths close to this shortest path?

5   Give a reason why a parabolic mirror brings light from a very distant source to a sharp focus.

6   If you 'filled in' the slits of a grating, why would the diffraction maxima vanish?

Answers 1. →, → **2.** →, ← **3.** they have phasors which are in phase **4.** they are nearly in phase **5.** the paths are all equal in length **6.** there would be extra paths with phasors cancelling those which give rise to the maxima

YOU HAVE LEARNED

- That photons will seem to go along special paths if near those paths the phasor arrows 'line up', whilst elsewhere they 'curl up'.

- That phasor arrows for photons line up for paths near the path which takes the least time.

- That these principles account for reflection, refraction, straight line propagation, help in the design of curved mirrors and lenses, and account for diffraction from a slit.

- That gratings get certain definite paths to have phasors which 'line up' by removing paths which when added in make the phasors 'curl up'.

quantum amplitude, many paths, reflection, refraction, diffraction

7.3 Electrons do it too

We said at the beginning of this chapter that quantum behaviour is important because it is common to all the fundamental particles you may read about, including neutrinos, quarks and others. Here we will extend the ideas to electrons.

An electron is not exactly like a photon. Electrons have mass, which means that they can travel at different speeds. Photons always travel at the speed of light.

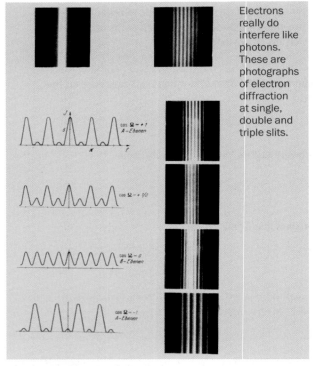

Electrons really do interfere like photons. These are photographs of electron diffraction at single, double and triple slits.

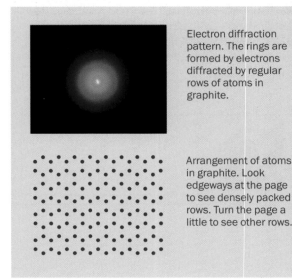

Electron diffraction pattern. The rings are formed by electrons diffracted by regular rows of atoms in graphite.

Arrangement of atoms in graphite. Look edgeways at the page to see densely packed rows. Turn the page a little to see other rows.

But, just like a photon, an electron offered more than one possible path explores them all. Just like a photon, a rotating arrow or phasor tracks the phase change along every path, and these arrows superpose in just the same way.

It took some ingenuity to repeat Young's classic two-slit experiment, to show that electrons can interfere just like photons. To see electron quantum behaviour, it is necessary to use atom-sized spacings of slits. Atoms arranged in regular rows can be used as a natural finely spaced grating.

It's easy to make the electrons go faster or slower by changing the accelerating potential difference. If you do, the diffraction pattern changes. The rate at which the phasor arrow for an electron path turns depends on the energy of the electrons.

How fast do electron arrows turn?

Richard Feynman made an inspired guess about how to calculate how fast the electron arrows turn following a suggestion made by Paul Dirac. For photons, the rate of phasor turning is just:

$$f = \frac{E}{h}$$

The guess was that for electrons moving with kinetic energy E_{kinetic} the frequency of turning might simply be

$$f = \frac{E_{\text{kinetic}}}{h}$$

If the electron has potential energy, this must be subtracted from the kinetic energy.

How far will electrons travel while making one turn—their wavelength? The answer was first suggested by Louis de Broglie when quantum ideas were first being developed. It is:

$$\lambda = \frac{h}{momentum}$$

The quantity momentum calculated at speeds much less than the speed of light is mv, the mass m of the electron multiplied by its velocity v. You can check it by varying the speed of electrons in an electron diffraction tube, through changing the accelerating potential difference.

- How 'Try all paths' predicts the wavelength for particles
- Energy, momentum and quantum behaviour
- The principle of least action

Try these

Charge on electron = -1.6×10^{-19} C, mass of electron = 0.9×10^{-30} kg, Planck constant $h = 6.6 \times 10^{-34}$ Js.

1 Electrons are accelerated by a potential difference of 100 V. What is their kinetic energy?

2 At what rate does the phasor turn for such an electron?

3 What is the speed of these electrons?

4 What is the de Broglie wavelength associated with these electrons? Comment.

5 What problem would there be in doing a two-slit interference experiment with such electrons?

6 What would be the speed and de Broglie wavelength of an electron with energy similar to that of a visible photon, say 1 eV?

Answers 1. 1.6×10^{-17} J **2.** 2.4×10^{16} Hz **3.** approximately 6×10^6 m s^{-1} **4.** 1.2×10^{-10} m, or about 0.1 nm; similar to the size of an atom **5.** the slit spacing would have to be only a few atoms wide **6.** 6×10^5 m s^{-1}, 1.2 nm

YOU HAVE LEARNED

- That electrons also show quantum behaviour.

- How to calculate the rate of rotation of the phasor for free electrons $f = E_{kinetic}/h$

- That the wavelength associated with electrons is $\lambda = h/mv$ (at low velocities).

- How to show that electrons produce superposition effects when passed through a grating.

electron diffraction, de Broglie wavelength

7.4 What does it all mean?

A peculiar thing about quantum behaviour is that although everyone agrees how to work things out, and everything agrees with experiment, there's more than one picture of what lies behind the calculations. There is no agreement about which is the 'right' picture. That's just how it is. So we have chosen the story which seems to us to be the simplest, a story first worked out by the physicist Richard Feynman. This was part of a development of quantum ideas around 1945–50 which won him, his fellow American Julian Schwinger and the Japanese physicist Sin-Itiro Tomonaga the Nobel prize. Tomonaga got there first, even in war-torn Japan.

You mustn't suppose that the picture we have painted is 'exactly how things are'. The story of photons following all possible paths is an easy way of remembering how to do the calculations. But don't let it carry you away. For example, don't imagine photons trying paths one at a time, one after the other. If anything, they must be imagined trying them all at once.

Feynman's 'try all paths' idea is not altogether new. Huygens (see chapter 6) long ago sowed its seeds. He thought of his wavelets as spreading out everywhere they could, and building up the new position of a wave by arriving there all in phase. The wavelets go everywhere, but in most places add up to nothing at all.

Shocking or not?

> *'Anyone who is not shocked by quantum theory has not fully understood it.'*
> Neils Bohr

People originally did find quantum behaviour very peculiar, even shocking. Many books have been written about its weirdness. You will have to decide for yourself, when you have got used to it.

It does seem peculiar that quantum behaviour steals the mathematics of waves, without supposing that waves lie behind it. But let us try to turn this question round: why is a wave theory of light possible, if photons are quantum objects?

Here's an answer. See what you think of it. Quantum behaviour is governed by combining

Richard Feynman (right). Feynman's life is vividly described in James Gleick's book *Genius*. Feynman was renowned for his lightning speed calculating ability and for his 'no nonsense' approach to problems in physics and in life.

phasors (amplitude and phase in one spinning arrow) to calculate probabilities. Suppose that this is *all there is to say* about how photons work. What happens when there are countless numbers of photons in a beam? The probability just determines the brightness, which varies smoothly from place to place. And the brightness varies just as if waves were superposing to give the result, *because of* the quantum adding-up of phasors.

This turns the usual argument on its head. The usual argument says that photon behaviour is calculated by wave-like methods because photons are a bit like waves. Our argument says that light is a bit like waves because photons show quantum behaviour. Waves are what photons appear to do when the photons are numerous.

The argument may feel uncomfortable. Something almost obvious is being explained in terms of something strange. Shouldn't it be the other way round? But if so, how will a familiar thing ever get explained?

Waves or particles?

The modern story we have told avoids the puzzle: 'Are these things waves or particles?'. Many a book about quantum physics scratches this ancient sore spot. Result: it just itches more. Our answer is that electrons and photons are neither—they are just themselves, 'quantum objects'.

This doesn't mean that the puzzles go away. Photons somehow manage both to go everywhere and yet always to be detected somewhere. Arguments about the consequences of this 'non-local' yet 'localised' behaviour continue to this day. They are leading to new ideas about 'quantum computing' which, if it can be made to work, would be immensely powerful.

Summary checkup

✓ Quantum behaviour:

- Quantum behaviour is unique; neither wave nor particle behaviour. Its fundamental rule is 'Try all paths'.

✓ Photons:

- Photons show quantum behaviour.

- The energy delivered by a photon is $E = hf$

- Quantum behaviour combines phasors from all possible paths.

- The probability of an event is found from the square of the resultant phasor amplitude.

✓ Paths taken by photons:

- Photons will seem to go along special paths if near those paths the phasor arrows 'line up', whilst elsewhere they 'curl up'.

- Phasor arrows for photons line up for paths near the path which takes the least time.

- These principles account for

    reflection
    refraction
    straight line propagation

  and

    help in the design of curved mirrors and lenses
    account for diffraction from a slit.

- Gratings get certain definite paths to have phasors which 'line up' by removing paths which when added in make the phasors 'curl up'.

✓ Electrons:

- Electrons also show quantum behaviour.

- The rate of rotation of the phasor for free electrons (zero potential energy) is $f = E_{kinetic}/h$.

- The wavelength associated with electrons is $\lambda = h/mv$ (at velocities much less than the speed of light).

- Electrons produce superposition effects when passed through a grating, usually on an atomic scale.

Questions

1 Paths for photons travelling from a source to a detector via a mirror are shown. Three paths go via A near one edge of the mirror. Three paths go via B where the angle of incidence is equal to the angle of reflection. The graph shows how the trip time varies with the position of the point where the path meets the mirror.

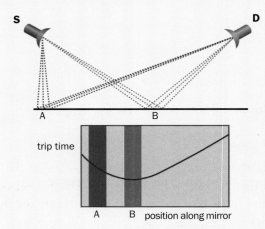

(a) Are the trip times for the paths near A similar?

(b) Are the trip times for the paths near B similar?

(c) Sketch the three phasors at the detector for the paths going near A.

(d) Sketch the three phasors at the detector for the paths going near B.

2 Fill in the empty spaces in the tables below. Planck constant $h = 6.6 \times 10^{-34}$ J s, speed of light $c = 3 \times 10^{8}$ m s^{-1}, charge on electron $e = 1.6 \times 10^{-19}$ C, mass of electron $= 9.1 \times 10^{-31}$ kg. Kinetic energy $= \frac{1}{2}mv^2$. Energy in joule = energy in electron volt × charge on electron.

Photons

frequency of phasor rotation	energy	wavelength
		300 nm
300 GHz		
	1.6×10^{-19} J	

Electrons

frequency of phasor rotation	energy	de Broglie wavelength
	10 electron volt	
		10 nm
10^{16} Hz		

3 Photons go from a source to a detector, as shown.

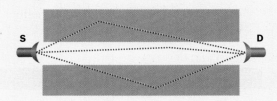

(a) Why do paths going through the shaded regions contribute little to the resultant phasor at the detector for all paths between source and detector?

(b) What has this to do with the rectilinear propagation of light?

4 Light goes through a single narrow slit. Many photons arrive at a place at an angle to the direction of the light falling on the slit.

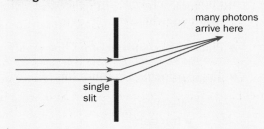

(a) Are the trip times of the three paths from the slit equal?

(b) If many photons arrive at the point shown, what can you say about the numbers of phasor rotations for the three paths?

5 A converging lens is focusing light, as shown.

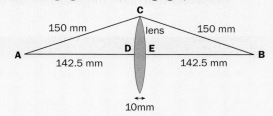

(a) Express the speed of light in mm s^{-1}.

(b) How long does the light take to follow the route ACB?

(c) How long does the light take to cover the route ADEB?

(d) How long must the light spend between D and E, in the glass?

(e) What is the speed of light in the glass?

(f) What is the refractive index of the glass?

8 Mapping space and time

Here we introduce you to a new kind of mathematical quantity—vectors. Vector quantities provide a powerful way to describe arrangements in space, movements, forces, fields and many other things. We will give examples of:

- vectors in map-making

- vectors describing movement

- how to take vectors apart into components

- how to add vectors together

These ideas are important in all parts of physics and engineering.

8.1 Journeys

Robert Reid is a physicist who has studied cosmic rays at the South Pole. He describes his journey to the Pole: "The flight took eight hours in a Hercules aircraft, sitting in a cramped hammock seat, surrounded by cargo, with no heating and no insulation from the sound of the engines. It was too dark to read and the only sanitation was a bucket." The time the journey took, not the distance he went, is what sticks in his mind. You have probably felt exactly the same on a long trip.

Time is often used to describe distance. Bats listen to echoes of high frequency chirps they make, to find out about objects in front of them. Submarines use the same idea. Electronic tape-measures from DIY stores use ultrasound reflections to measure rooms; medical ultrasound scans locate babies in the womb (chapter 1).

Robert Reid throws his favourite omega-shaped boomerang around the South Pole (marked by the board). This boomerang has a circular orbit of diameter 40 m and returned accurately directly behind the thrower. Time of flight? Choose for yourself from 9 seconds or 1 day and 9 seconds!

To get from the journey time to the distance, you need to know the speed. If you can guess the speed of a Hercules aircraft, you can estimate the length of Robert Reid's trip. But the idea is deeper than you might think. The speed of light is special: it doesn't depend on how you are moving. So all journeys really *can* be described by times, the time light would take to make them. This is the starting point for Einstein's theory of relativity.

Transport

Over the past hundred years, transport has developed so quickly that the distances we travel every day would have been unthinkable a century ago. If we think about a day's travel (say eight hours time on the road) as a sensible unit for measuring journey times, how has the distance that we can travel in this time changed?

Today, we can travel huge distances in a day. The geneticist Steve Jones has written, "...there is little doubt that the most important event in recent human evolution was the invention of the bicycle"; according to Steve Jones it allowed mixing of the gene pool in ways that had not happened before its invention, when people typically married others in their own or very close communities. Transport has given people enormous personal freedom.

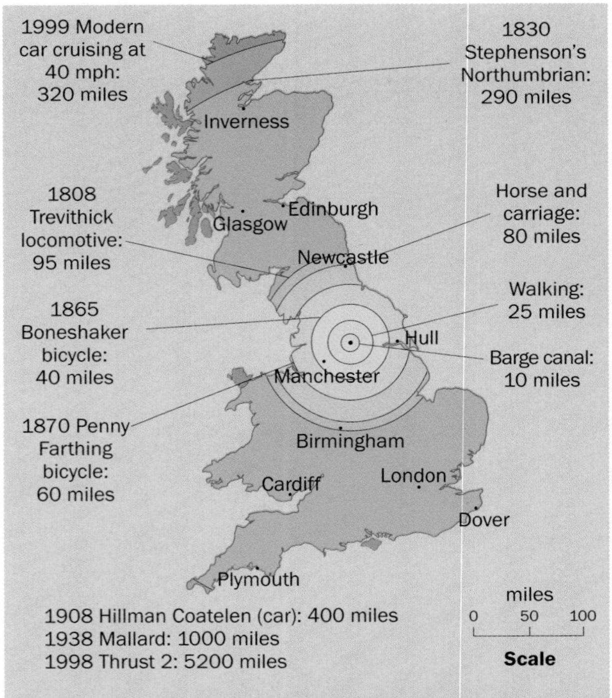
1999 Modern car cruising at 40 mph: 320 miles
1830 Stephenson's Northumbrian: 290 miles
Inverness
1808 Trevithick locomotive: 95 miles
Edinburgh
Glasgow
Horse and carriage: 80 miles
Newcastle
Walking: 25 miles
1865 Boneshaker bicycle: 40 miles
Hull
Barge canal: 10 miles
Manchester
1870 Penny Farthing bicycle: 60 miles
Birmingham
Cardiff
London
Dover
Plymouth
1908 Hillman Coatelen (car): 400 miles
1938 Mallard: 1000 miles
1998 Thrust 2: 5200 miles
miles
0
50
100
Scale

How far can you travel from Leeds in eight hours? The circles show the distances that could be travelled at the constant top speeds of these types of transport in eight hours. Why are these distances an overestimate?

Speed

The route from Leeds to Manchester through the Pennine hills of the north of England has long been important for trade; it has carried a packhorse track, a canal, a railway and a motorway. The straight line distance from Leeds to Manchester is about 35 miles; it takes the cross-country train an hour to travel this distance, a car about 75 minutes and a barge about 33 hours. In each case, the average speed for the journey is somewhat less than you might expect from an estimate of the speed of the type of transport. In a car, I can drive at 70 miles per hour along the motorway, so that the time to travel 35 miles is:

$$\text{speed} = \frac{\text{distance}}{\text{time}} \quad \text{or in symbols } v = \frac{s}{t}$$

$$\text{so time} = \frac{\text{distance}}{\text{speed}}$$

$$\text{or time} = \frac{35 \text{ miles}}{70 \text{ miles hour}^{-1}} = 0.5 \text{ hours}$$

Of course, the actual time is much longer than this because I cannot drive at this speed all the way: city centre traffic soon sees to this, as does rush hour traffic on the M62. A distance–time graph makes this clear.

This graph tells a story. Think about what is happening at each stage of the journey. Where was

A journey from Leeds to Manchester

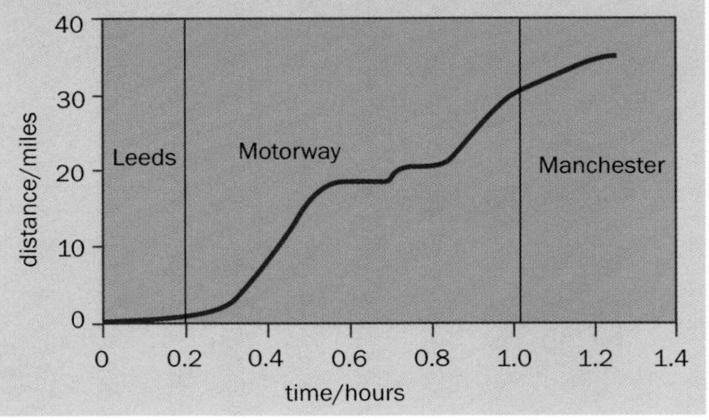
40
30
20
10
0
distance/miles
Leeds
Motorway
Manchester
0
0.2
0.4
0.6
0.8
1.0
1.2
1.4
time/hours

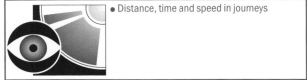

• Distance, time and speed in journeys

I stuck? Where did I have to slow down? In a sense, this graph is a map of the journey. A conventional road map lays out the route in two space dimensions. This map lays out the journey in a two-dimensional 'space-time'. It is a frozen story of the journey.

Instantaneous speed and average speed

At times, travelling from Leeds to Manchester in the car, I might drive at 70 mph, but at other times I sit gridlocked. What is my average speed? Easy! This is just the total distance divided by the total time:

$$\text{average speed} = \frac{35 \text{ miles}}{1.25 \text{ hours}}$$

$$= 28 \text{ mph or } 28 \text{ miles hour}^{-1}$$

Here is something odd: the speedometer in my car swings round to 70 mph as I move away from the traffic jam, but it does not need to wait until I have travelled for an hour to calculate this: so what is it reading? It tells me the *instantaneous* speed of the car, the speed *at that moment*. In practice, I could estimate this by seeing how far the car travels in a very short space of time: this is what light gates do that you may use in practical work. The shorter the time interval, the more accurate is the approximation to the instantaneous speed.

Let's look at a graph of my journey along the motorway and see what average and instantaneous speeds mean.

The speeds—both average and instantaneous—can be calculated from the *gradient* of a line. Notice that because the *gradient* is calculated by dividing a distance by a time its unit must be that of a speed.

Have a look at the image of a woman walking. Her hand and her head must have the same average speed (why?) but what about their instantaneous speeds at given moments?

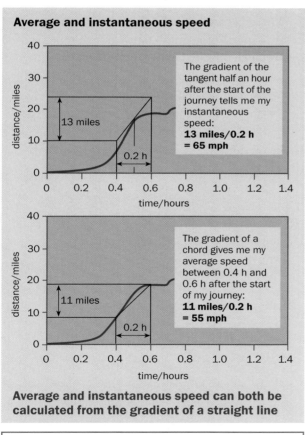

Average and instantaneous speed

The gradient of the tangent half an hour after the start of the journey tells me my instantaneous speed:
13 miles/0.2 h = 65 mph

13 miles

0.2 h

The gradient of a chord gives me my average speed between 0.4 h and 0.6 h after the start of my journey:
11 miles/0.2 h = 55 mph

11 miles

0.2 h

Average and instantaneous speed can both be calculated from the gradient of a straight line

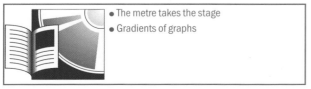

- The metre takes the stage
- Gradients of graphs

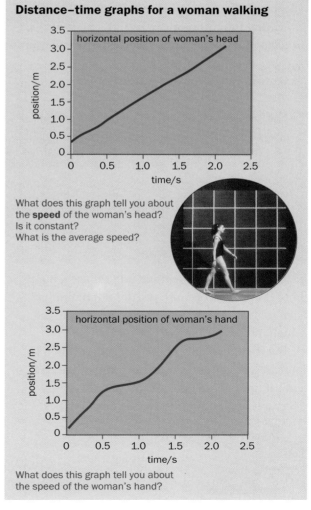

Distance–time graphs for a woman walking

horizontal position of woman's head

What does this graph tell you about the **speed** of the woman's head? Is it constant? What is the average speed?

horizontal position of woman's hand

What does this graph tell you about the speed of the woman's hand?

Speed–time graphs

Suppose on my journey from Leeds to Manchester along the M62 I am able to travel at a steady speed of 70 mph for 30 minutes. This time, let's convert to SI units: 70 mph is roughly 30 m s^{-1} and 30 minutes is 1800 s. How far will I have travelled?

$$\text{speed} = \frac{\text{distance}}{\text{time}}$$

so

distance = speed × time
distance = 30 m s^{-1} × 1800 s
distance = 54 km

You've already thought about one way of picturing a journey: as a distance–time graph. Now look at a different picture: a speed–time graph.

In this example, the graph is very simple. Look at the shaded part, a rectangle whose 'area' is 30 m s^{-1} multiplied by 1800 s. This is just the quantity calculated above, the distance travelled. The graph shows that this distance can be thought of as the area below a graph. Notice the rather peculiar units

downwards. The speed–time graph reaching zero means that you have stopped.

It's easy to see that the rectangular area under the graph for constant speed is the distance travelled, because distance = speed × time. But it isn't so easy to see that the area under *any* speed–time graph is the distance travelled, no matter how the speed changes.

One way to see it is to think of a very short interval of time, say a few seconds. In that short time the speed can't change much, so there is a more-or-less definite speed v for that moment. The distance gone in the short time interval Δt is $v\Delta t$. This too is an area beneath the curve: it is the area of a thin strip, Δt wide, reaching up a height v to the curve.

Now imagine doing the same, strip after strip across the whole curve. The total area is the area of all the strips added together. The total distance travelled is the area under the curve.

This gives a very useful general rule: the area below a speed-time graph between two times is the distance travelled between those times.

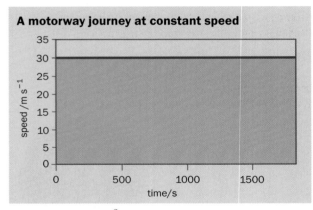

A motorway journey at constant speed

of the area: not m^2, the usual unit for area, but m. This of course is because you are not multiplying a length by a length but rather a speed by a time. Watch out!

But the speed may not be constant. When it increases, as you speed up, the graph of speed against time slopes upwards. When the speed decreases, as you slow down, the graph slopes

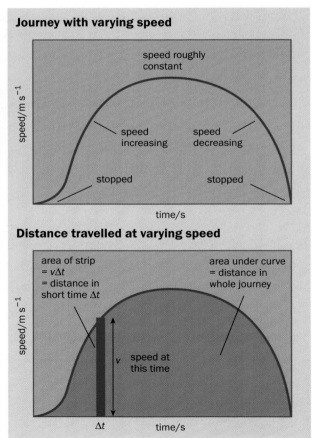

Journey with varying speed

speed roughly constant

speed increasing

speed decreasing

stopped

stopped

time/s

Distance travelled at varying speed

area of strip
= $v\Delta t$
= distance in short time Δt

area under curve
= distance in whole journey

v speed at this time

Δt time/s

- Travel graphs with Modellus
- Investigating motion graphs with the Multimedia Motion CD-ROM

Try these

1 A Hercules aircraft flies at a speed of 250 m s^{-1}. How far can it fly in 8 hours?

2 I drive at 70 mph for 30 minutes and at 30 mph for another hour. How far do I travel altogether? What was my average speed? Is it the same as the average of the numbers 30 and 70?

3 Draw a distance–time graph for the journey of question 2. Draw a speed–time graph for this journey. Annotate the distance–time graph to show how the speed can be calculated. Annotate the speed–time graph to show how the distance can be calculated.

4 At an average speed of 2 m s^{-1}, how far could I walk in 8 hours?

5 The speed limit in built-up urban areas is 30 mph. What is this in m s^{-1}?

6 Look at the graph of my journey from Leeds to Manchester on page 178. What is my average speed as I travel from Leeds to Manchester? What is my instantaneous speed an hour after the start of my journey?

7 By thinking about the gradient of the graph, use the distance–time graph of my journey from Leeds to Manchester to sketch a speed–time graph. Annotate it with comments about what you consider to be the important features.

Answers 1. 7200 km **2.** 65 miles; 43 mph; no **4.** 57.6 km **5.** 13.5 m s^{-1} **6.** about 28 mph; about 25 mph

YOU HAVE LEARNED

- That speed is distance over time; $v = \dfrac{s}{t}$.

- That speed can be calculated from the gradient of a distance–time graph.

- The meanings of average and instantaneous speeds and how these are calculated from speed–time graphs.

- The meaning of the area under a speed–time graph and how to estimate this.

speed, average speed, instantaneous speed, gradient, tangent

Table of distances

	Bordeaux	Calais	Lyon	Marseille	Paris	Brest
Bordeaux						
Calais	696					
Lyon	426	612				
Marseille	498	879	270			
Paris	492	228	390	651		
Brest	489	534	759	939	486	

Distances in km taken along straight lines between places

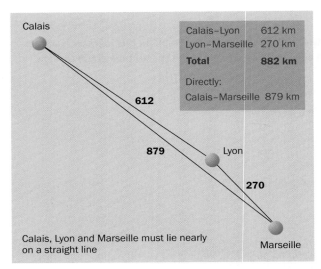

Calais–Lyon	612 km
Lyon–Marseille	270 km
Total	**882 km**
Directly:	
Calais–Marseille	879 km

Calais, Lyon and Marseille must lie nearly on a straight line

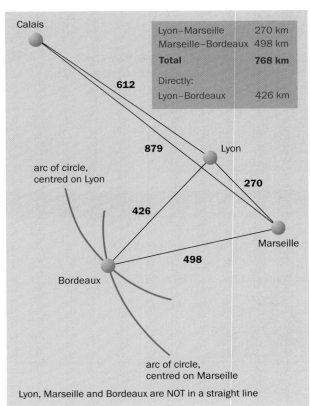

Lyon–Marseille	270 km
Marseille–Bordeaux	498 km
Total	**768 km**
Directly:	
Lyon–Bordeaux	426 km

arc of circle, centred on Lyon

arc of circle, centred on Marseille

Lyon, Marseille and Bordeaux are NOT in a straight line

8.2 Maps and vectors

Only on a circular walking trip do you set out to cover a big distance without going anywhere. Usually on a journey you want to go somewhere, to go a certain distance in a particular direction. Distance in a certain direction is called *displacement*, and it is a vector. *Vector quantities*, like my journey, have both size, usually called magnitude (distance in this case), *and* direction. Vectors are very important in physics: examples include velocity, acceleration and force, which we shall consider later in this chapter and in chapter 9. Quantities like speed which have only a magnitude are called *scalars*. Volume is another example of a scalar.

Suppose that you are on holiday in France. You don't know the country at all, but you have a table of the distances between towns ripped out of a road atlas. What the table does not tell you is the direction from one town to another. The table tells you the *scalar*, distance, not the *vector*, displacement. Curiously, it is possible to find the *relative* directions just from the distances. Seeing how to do this helps to understand how vector quantities combine.

Look at the table of distances for some of these French towns. We'll consider the shaded distances.

It is obvious from the figures, if you know what to look for—even if you know nothing about the map of France—that Lyon lies very nearly on a straight line from Calais to Marseille.

Not every pair of distances works out like this. Look at the figures for Marseille, Bordeaux and Lyon. These distances do not add up! The paths must be angled.

To find a possible layout of these towns, lines proportional to 426 km and to 498 km can be anchored on Lyon and Marseille, and rotated until they meet. (Notice that the location of Bordeaux is not the only possible one—the circular arcs would meet just as well above and to the right of Lyon and Marseille.)

What has this to do with vectors? A little notation will help. Call the distance from Bordeaux to Lyon *a*, and the distance from Lyon to Calais *b*. Finally, call the distance from Bordeaux to Calais *c*. If I write

$$a + b = c$$

then you may tell me I am wrong. But there is a sense in which this is correct. If **a** now means the journey from Bordeaux to Lyon (I have used a bold typeface, so that you can tell the difference between this and the simple distance), and similarly for **b** and **c**, then

a + **b** = **c**

does make sense: it means if I travel first from Bordeaux to Lyon and then from Lyon to Calais, I will end up in the same place as if I simply went from Bordeaux to Calais. This is how vectors add together. I have added the *displacement* from Bordeaux to Lyon to the *displacement* from Lyon to Calais to get the *displacement* from Bordeaux to Calais.

Bordeaux was placed where the two circular arcs centred on Lyon and Marseille met. But a check on that can be made, because we also know the distance from Bordeaux to Calais, and this is done in the picture.

Now that you have your map you can journey with confidence from one place to another—but wait a moment! You don't know which way up the map goes! Which way is north? In fact, the relative positions of the towns are independent of which way up the map goes: printing maps with north at the top is simply a convention. Notice that vector equations like **a** + **b** = **c** do not depend on the direction of north either. Choosing a special direction is important when dealing with vector components, which you will read about next.

Resolving vectors

Let's ask a crazy question. If I drive from Marseille to Bordeaux, how far towards Calais have I travelled? Since I was not heading the right way, I have gone 498 km and have 696 km left to go in a different direction.

But, another answer is that I have driven about 300 km along the true direction Marseille–Calais. That does not sound very sensible, since at Bordeaux I am nowhere near Lyon, having gone more than 400 km west as well. It sounds a bit more sensible if we ask how far north I have gone from Marseille. Then (if Calais is due north of Marseille, which is roughly true) the answer is about 300 km due north. You can see this in the picture, where the map has been rotated so that it is aligned conventionally, with north at the top.

The distance I have journeyed north is a *component* of the displacement vector from Marseille to

Bordeaux. The displacement from Marseille to Bordeaux can be *resolved* into two components: 300 km north and 400 km west (both distances are approximate).

But how can you find the components of a vector in given directions? One way of doing this is as we have on the map. We *projected* the journey from Marseille to Bordeaux onto the direction from Marseille to Calais. You can do this by scale drawing, or by using trigonometry.

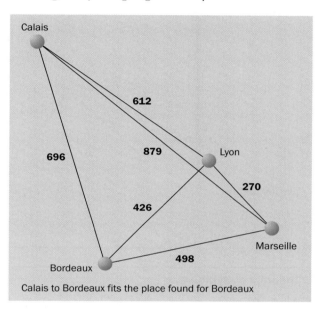

Calais to Bordeaux fits the place found for Bordeaux

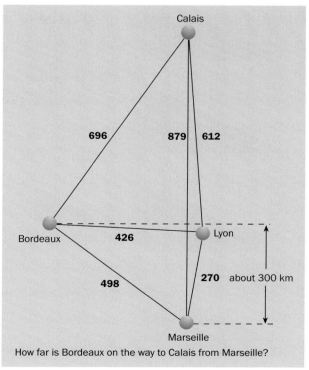

How far is Bordeaux on the way to Calais from Marseille?

Vectors

Notation

To show that quantities are vectors, three notations are used:

a – bold typeface, usual in printed text;

a – straight or wavy underline – the one to use in handwriting;

$\vec{a}$ – arrow over the letter – often used to indicate a vector from A to B, as $\overrightarrow{AB}$

To show the magnitude of a vector, the same letter is used either in a plain typeface or by using modulus signs: *a* or |*a*| or |_a_| or |$\vec{a}$|.

Take 2 vectors

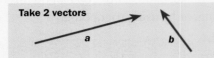

$$c = a + b$$

...move the tail of one to the tip of the other...

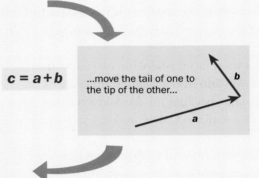

...and the result is drawn from the tail of the first to the tip of the second...

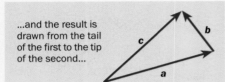

Vector addition

Vectors add by placing them 'tip to tail' and drawing the *resultant* vector from the tail of the first to the tip of the second. The *resultant* is the sum of the two vectors. In practice, finding a resultant can be done by scale drawing (as it is in a road atlas!)

Theorem of Pythagoras

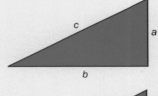

To prove:

$$c^2 = a^2 + b^2$$

Draw a perpendicular to the longest side, which divides it into two parts length *x* and *y*.

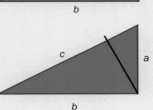

Turn and flip the two new triangles to line up with the first one.

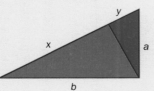

$$y/a = a/c$$
$$y = a^2/c$$

The two new triangles are both the same shape as the first one, but scaled down in size.

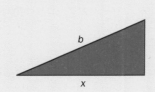

$$x/b = b/c$$
$$x = b^2/c$$

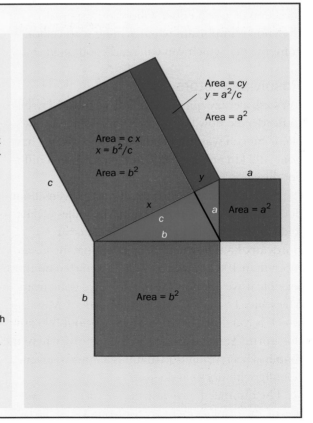

Area = cy
y = a²/c
Area = a^2

Area = c x
x = b²/c
Area = b^2

Area = a^2

Area = b^2

Components of a vector

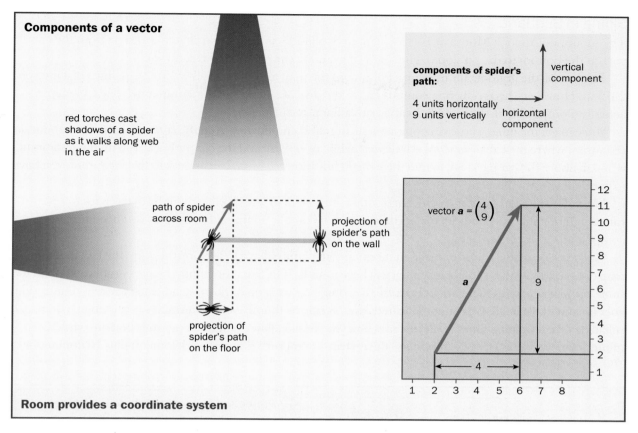

red torches cast
shadows of a spider
as it walks along web
in the air

components of spider's path:

4 units horizontally
9 units vertically

vertical component

horizontal component

path of spider
across room

projection of
spider's path
on the wall

projection of
spider's path
on the floor

vector $\mathbf{a} = \begin{pmatrix} 4 \\ 9 \end{pmatrix}$

a

9

4

Room provides a coordinate system

Components of a vector from angles

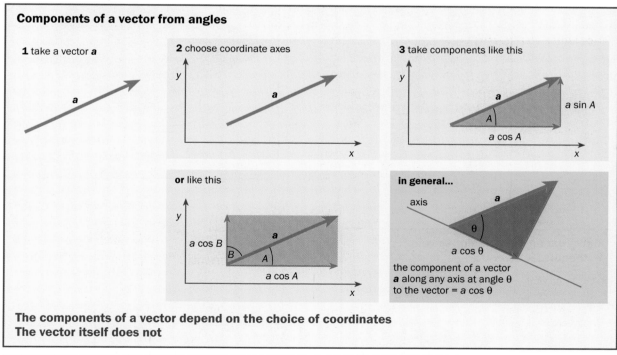

1 take a vector $\mathbf{a}$

a

2 choose coordinate axes

y

a

x

3 take components like this

y

a

$a \sin A$

A

$a \cos A$

x

or like this

y

$a \cos B$

a

B A

$a \cos A$

x

in general...

axis

a

θ

$a \cos \theta$

the component of a vector
$\mathbf{a}$ along any axis at angle θ
to the vector = $a \cos \theta$

The components of a vector depend on the choice of coordinates
The vector itself does not

- Adding vectors

- Understanding vector components
- Rotating coordinate systems

Skiing in Scotland

The mountain Aonach Mor near Fort William in Scotland has a cable car that takes skiers from the car park at the bottom to the ski slopes. How far is it from the car park to the restaurant?

The line of the gondola lift measures 95 mm on the map and the scale is 1:25 000, giving a distance of 2375 m. However, this is only one component of the displacement: the horizontal one! The OS map is actually a projection of the real landscape onto a flat piece of paper.

You can get the other, vertical, component of the cable car journey from the contour lines. The mountain restaurant where the cable car stops is built at a height of 660 m, and the car park is at 100 m: the restaurant is 560 m above the car park. So, to find the straight line distance from the car park to the restaurant, you have to use Pythagoras's theorem.

Just as for the map of France, the horizontal component of the displacement from car park to restaurant could be split into components north and east. Generally, in three dimensions, you need three components to describe vectors.

Walkers calculating times for walks in mountainous country have to take into account the fact that the map does not show the true distance they will have to walk. The situation is much more complicated than the cable car journey we thought about because they don't just walk up or down steady inclines: real countryside goes up and down in all sorts of complicated ways. A rule of thumb called Naismith's rule is often used to calculate rough walking times. This rule says that you should allow 20 minutes per mile and an extra 30 minutes for every 1000 feet of climbing—the metric version is 12 minutes per kilometre plus 10 minutes per 100 m of climbing.

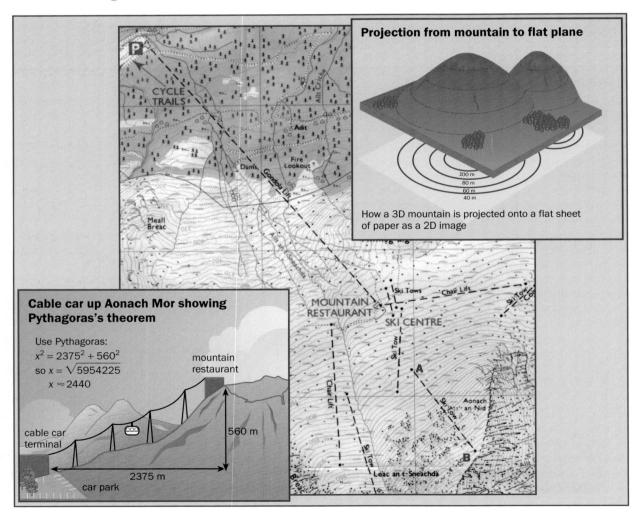

Projection from mountain to flat plane

100 m
80 m
60 m
40 m

How a 3D mountain is projected onto a flat sheet of paper as a 2D image

Cable car up Aonach Mor showing Pythagoras's theorem

Use Pythagoras:
$$x^2 = 2375^2 + 560^2$$
so $x = \sqrt{5954225}$
$$x \approx 2440$$

mountain restaurant

cable car terminal

560 m

2375 m

car park

Try these

1 Use Pythagoras's theorem to calculate how far I am from my starting point if I first drive 30 miles west and then 40 miles north.

2 Draw a scale drawing to show my total displacement if I first walk 200 m NW and then 300 m south. Use a scale of 1 cm to 20 m. How far am I from my starting point? In what direction? (Give the direction as a bearing or as an angle clockwise from north.)

3 An aeroplane flies 200 miles SE. By scale drawing, find how far south it travels from its starting point and how far east. (These are the components of the motion of the displacements along these directions.) Use a scale of 1 cm to 20 miles.

4 What is the vertical component of the displacement from A to B at the foot and the top of one of the ski tows marked on the Aonach Mor map? What are the horizontal components north and east?

5 Find the straight line distance between A and B on the ski tow.

6 Fort William in Scotland is just about 100 miles north west of Edinburgh. Use trigonometry to calculate how far north Fort William is from Edinburgh. How far west is Fort William from Edinburgh?

7 Use Pythagoras's theorem to check your answers to question 6 by verifying that the components give the correct distance from Edinburgh to Fort William.

Answers 1. 50 miles **2.** 212 m on bearing of 222° **3.** 141 miles south and 141 miles east **4.** 300 m; −550 m north, 425 m east **5.** approx. 760 m **6.** 71 miles north and 71 miles west

---

YOU HAVE LEARNED

- That vectors have magnitude and direction.
- That vectors are added by placing them tip to tail and drawing a resultant vector.
- The notation used to denote vectors.
- That vectors can be resolved into components at right angles.
- How to use Pythagoras's theorem to find the magnitude of a resultant vector when two vectors at right angles are added.

- How to use Pythagoras's theorem to calculate the magnitude of a vector from its components in two directions at right angles.
- How to use trigonometry to resolve a vector into components at right angles.

vectors, vector components, vector addition, scalar quantity, displacement

8.3 Velocity

The only vector we have discussed so far is displacement, distance in a certain direction. But if you set off on a journey, you want to know how long it is going to take. The vector related to speed in the same way that displacement is related to distance is *velocity*. Velocity means speed in a certain direction. To calculate velocity, you use the equation

$$\text{velocity} = \frac{\text{displacement}}{\text{time}} \quad \text{or in symbols} \quad \boldsymbol{v} = \frac{\boldsymbol{s}}{t}$$

Because to tell you your velocity, we have to be able to give you a direction as well as a speed and because we don't know when you are reading this, we don't know the whereabouts of the Earth in its orbit! Are your average and instantaneous velocities the same? No! The direction in which you are moving is changing second by second as the Earth moves around its orbit. Notice how careful you have to be when you talk about vectors: here, there is a constant speed, a velocity which changes second by second and an average velocity which is equal to zero!

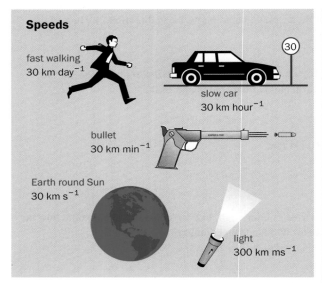

Speeds

fast walking
30 km day^{-1}

slow car
30 km hour^{-1}

bullet
30 km min^{-1}

Earth round Sun
30 km s^{-1}

light
300 km ms^{-1}

Notice that you have to make sure both sides of the equation are vectors. Velocity and displacement are represented by bold letters, $\boldsymbol{v}$ and $\boldsymbol{s}$, but time, a scalar, is represented by a plain typeface t. Notice that dividing a *vector* by a *scalar* on the right hand side of the equation gives you a vector.

Just as we can have an average speed and an instantaneous speed, so we can talk about average velocity and instantaneous velocity. Whilst reading this book, you are on a journey, moving at a speed of about 30 kilometres per second. Why? Because the Earth is rotating about the Sun (note that we have ignored the spin of the Earth about its axis, the rotation of the Galaxy and so on!). What about your velocity? We don't know what that is! Why not?

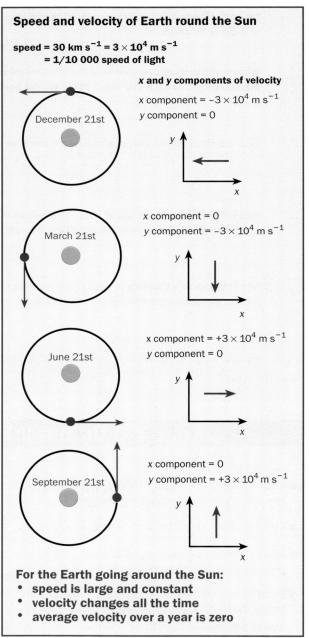

Speed and velocity of Earth round the Sun

speed = 30 km s^{-1} = 3 × 10^4 m s^{-1}
= 1/10 000 speed of light

x and y components of velocity

December 21st
x component = −3 × 10^4 m s^{-1}
y component = 0

March 21st
x component = 0
y component = −3 × 10^4 m s^{-1}

June 21st
x component = +3 × 10^4 m s^{-1}
y component = 0

September 21st
x component = 0
y component = +3 × 10^4 m s^{-1}

For the Earth going around the Sun:
• **speed is large and constant**
• **velocity changes all the time**
• **average velocity over a year is zero**

• Getting a feel for velocity vectors

Adding velocities

If you have ever flown across the Atlantic, you will know that the time to fly from west to east is less than the time it takes to fly from east to west. This is because of the Jet Stream, a current of air that blows more or less steadily across the Atlantic from west to east. It is produced by convection currents in the atmosphere, between the warm tropics and the cold north polar region. The aeroplane flies through the air at the same speed in both directions, but not at the same speed over the ground. This is because the air is moving too. To find the plane's

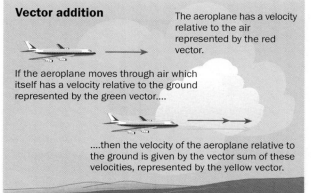

Vector addition

The aeroplane has a velocity relative to the air represented by the red vector.

If the aeroplane moves through air which itself has a velocity relative to the ground represented by the green vector....

....then the velocity of the aeroplane relative to the ground is given by the vector sum of these velocities, represented by the yellow vector.

velocity relative to the ground we have to add together the velocity of the air and the velocity of the aeroplane.

Suppose that as the pilot flies from Kennedy Airport, New York, to Heathrow, London the plane has to cope with a wind blowing from north to south across its flight path. What happens if the pilot simply ignores this and tries to fly on a line west to east? Adding the plane's velocity vector to that of the wind, the plane flies, relative to the ground, in the wrong direction! To find out which way it is flying, you can use trigonometry.

Which way should the pilot fly the plane, relative to the air, in order to travel west to east relative to the ground? The plane can only fly at a speed of 300 m s^{-1} relative to the air, so it has to fly at an angle so that the combined effect of the wind velocity and the aeroplane's velocity is in the right direction.

- Vector addition in one dimension
- A boat crossing a river

Components of velocity

Just as you can think about the components of a displacement, velocity too has components. Imagine taking a video recording of someone walking diagonally across the room in front of you. What do you see when you watch your film-making efforts? On the TV screen, you see someone moving across the screen from right to left: you don't see the forward motion of the walker except as an increase in their size! What you see is only one component of their motion, and if you calculate their speed from right to left, you have, in fact, calculated only one

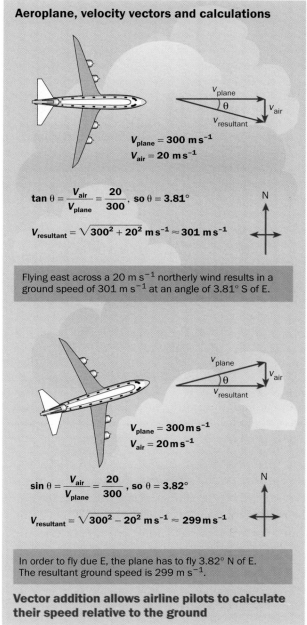

Aeroplane, velocity vectors and calculations

$V_{plane} = 300 \text{ m s}^{-1}$
$V_{air} = 20 \text{ m s}^{-1}$

$$\tan\theta = \frac{V_{air}}{V_{plane}} = \frac{20}{300}, \text{ so } \theta = 3.81°$$

$$V_{resultant} = \sqrt{300^2 + 20^2} \text{ m s}^{-1} \approx 301 \text{ m s}^{-1}$$

Flying east across a 20 m s^{-1} northerly wind results in a ground speed of 301 m s^{-1} at an angle of 3.81° S of E.

$V_{plane} = 300 \text{ m s}^{-1}$
$V_{air} = 20 \text{ m s}^{-1}$

$$\sin\theta = \frac{V_{air}}{V_{plane}} = \frac{20}{300}, \text{ so } \theta = 3.82°$$

$$V_{resultant} = \sqrt{300^2 - 20^2} \text{ m s}^{-1} \approx 299 \text{ m s}^{-1}$$

In order to fly due E, the plane has to fly 3.82° N of E. The resultant ground speed is 299 m s^{-1}.

Vector addition allows airline pilots to calculate their speed relative to the ground

component of their velocity. Their actual motion is *projected* onto a flat screen just as the OS map projects rugged terrain onto a flat piece of paper. (This is not quite true: if it were, the size of the person would not change!) Of course, if you had taken your video from above, you would have been able to see the true motion.

Think again about the journey of the Earth and other bodies through space: when you look at the night sky you can't tell how far away the stars are. It is easy to understand why it was once thought that the stars were holes in a heavenly sphere. But, of course, stars are not holes and they move through space. Every night, you can watch the constellations move round the sky because of the rotation of the Earth, but over much longer periods of time it is possible to detect that some stars have moved relative to their neighbours. The motion seen is only one component of the motion of the stars. You cannot see the motion of stars towards or away from you although this motion can in fact be detected by changes in spectral lines. Only the projection of the motion onto the heavenly sphere, the two components of the motion at right angles to your line of sight, is visible as a change in position.

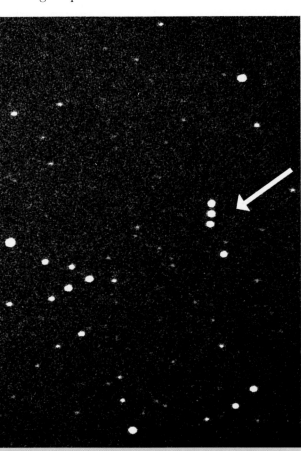

A visibly moving star (Barnard's star). The three arrowed images are of Barnard's star, photographed three times at two year intervals and superimposed. It has moved against the background of other stars. Of course only the components of its motion at right angles to the line of sight are seen.

The video stills on the left show only the motion perpendicular to the camera's line of sight, if you ignore the increase in size of the person.

The video clip of a falling chimney shows both vertical and horizontal motion. Might there be a component of motion that is not visible in the video?

• Distances fallen by a chimney

Vectors everywhere

You will find vector quantities appearing again and again in physics and engineering, from the design of bridges to the study of ocean currents; from the motion of galaxies in the Universe to the building of a dynamo.

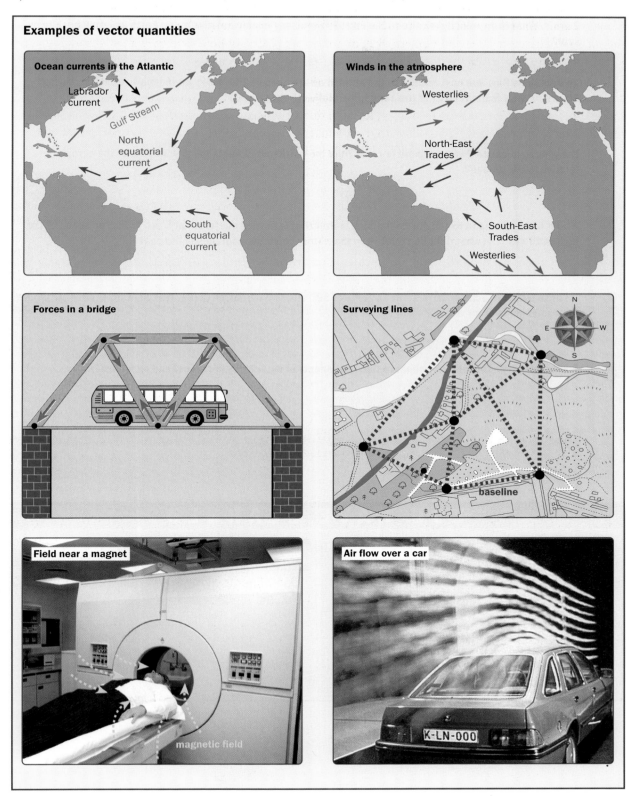

Examples of vector quantities

Ocean currents in the Atlantic

Labrador current

Gulf Stream

North equatorial current

South equatorial current

Winds in the atmosphere

Westerlies

North-East Trades

South-East Trades

Westerlies

Forces in a bridge

Surveying lines

baseline

Field near a magnet

magnetic field

Air flow over a car

K-LN-000

Try these

1 Whilst on a train travelling north at 50 m s^{-1}, I get up and walk towards the buffet car at the rear of the train at 2 m s^{-1}. What is my velocity relative to the Earth? What is my velocity relative to the Earth as I walk back to my seat?

2 An aeroplane flies due east relative to the air with an airspeed of 250 m s^{-1}. A wind blows from the north at 20 m s^{-1}. What is the velocity of the aeroplane relative to the ground?

3 The wind in question 2 veers around to come from the northeast. What is the new velocity of the aeroplane relative to the ground?

4 A boat has to be made to sail due east across a river that is flowing due south at 5 m s^{-1}. The top speed of the boat is 10 m s^{-1}. In what direction relative to the water should the captain sail her boat?

5 If I drive with a velocity of 20 m s^{-1} at an angle of 30° clockwise from north (i.e. on a bearing of 030°) what are the components of my velocity north and east?

6 For the velocity of question 5, what are the components of my velocity northwest and northeast?

Answers 1. 48 m s^{-1}; 52 m s^{-1} **2.** 251 m s^{-1} on a bearing of 094.5° **3.** 236 m s^{-1} on a bearing of 093.4° **4.** on a bearing of 060°
5. 17.3 m s^{-1} north, 10 m s^{-1} east **6.** 5.2 m s^{-1} northwest, 19.3 m s^{-1} northeast

YOU HAVE LEARNED

- That velocity is a vector.

- That velocity $= \dfrac{\text{displacement}}{\text{time}}$; $v = \dfrac{s}{t}$.

- About average and instantaneous velocity.

- That velocities can be resolved into components along different directions.

A-Z velocity, average velocity, instantaneous velocity; vector components; resolution of vectors; area under a curve

Summary checkup

✓ Motion:

- Speed $= \dfrac{\text{distance travelled}}{\text{time taken}}$; $v = \dfrac{s}{t}$
- The gradient of a distance–time graph gives speed.
- The area under a speed–time graph gives distance travelled.
- Distance–time and speed–time graphs can give alternative views of the same journey.
- Both distance–time and speed–time graphs can be interpreted to give a description of a journey.
- The average speed is the total distance divided by the total time.
- The instantaneous speed is the speed at a given moment, equal to the gradient of a distance–time graph.

✓ Vectors:

- Scalar quantities have no direction; vector quantities are fully specified only if a magnitude and a direction are given (or two components).
- Distance and speed are scalar quantities, displacement and velocity are the corresponding vectors.
- Vectors add geometrically, that is by placing them tip to tail, to form a resultant vector.
- Vectors can be added algebraically; special cases are:

    The resultant is the sum of the magnitudes if the vectors are in the same direction;
    The resultant is the difference between the magnitudes if the vectors are in opposite directions;
    The resultant is given by Pythagoras's theorem if the vectors are at right angles.

- The length of a vector is independent of a coordinate system but its components are not.
- Vectors may be resolved into components at right angles.

Questions

1 Below is a distance–time graph showing the height of a stone dropped from a tall building.

A stone falling from a tall building

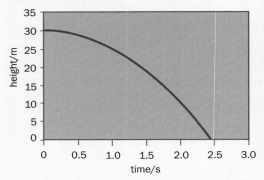

(a) What is the height of the building?

(b) At what speed was the stone dropped? Explain your answer.

(c) Use the graph to estimate the speed of the stone as it hit the ground.

(d) Use the graph to calculate the average speed of the stone during its fall. How does this compare with

   (i) the speed of the stone after half the time between release and hitting the ground has passed;

   (ii) the speed of the stone after the stone has fallen half the height of the building;

   (iii) the average of the speed at the top and the speed at the bottom of the fall.

(e) Draw a speed–time graph of the fall of the stone.

2 The table gives distances 'as the crow flies' in km between a cluster of towns.

	Leeds	Huddersfield	Batley	Bradford	Bingley
Huddersfield	22.5				
Batley	11.8	10.7			
Bradford	13.2	16.2	10.5		
Bingley	19.7	22.2	18.5	8.2	
Wakefield	13.0	19.2	12.0	20.7	28.7

(a) Without making a drawing, give a reason to think that Leeds, Batley and Huddersfield lie along a straight line.

(b) Still without drawing, give a reason to think that Leeds, Bradford and Huddersfield do **not** lie on a straight line.

(c) Use the distances to construct a possible map of the layout of these towns.

(d) Turn or flip your map to produce an alternative map. What does **not** change as you do so?

3 The graph shows part of the journey of an intercity train.

Part of the journey of an intercity train

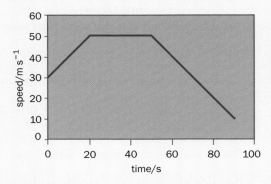

(a) Write a few sentences to describe the journey. In what ways is the graph unrealistic?

(b) How far did the train travel during the first 20 s shown on the graph?

(c) How far did the train travel during the 90 s shown on the graph?

(d) Draw a distance–time graph of the journey.

4 A girl can swim at 1.2 m s^{-1}. She swims across a river flowing at 0.5 m s^{-1}.

(a) If she swims perpendicularly to the flow of the water, what is her velocity relative to the bank?

(b) If the river is 10 m wide, how long would it take her to swim across the river?

(c) In what direction relative to the water would she have to swim in order to move perpendicularly to the bank?

(d) In the situation described in part (c), how long would it take her to swim across the river?

5 The following speed–time data were taken for a car.

Time/s									
0	1	2	3	4	5	6	7	8	9

Speed/m s^{-1}									
0	3.2	6.4	9.4	12.5	15.4	18.0	20.0	20.8	20.8

(a) Draw a speed–time graph for these data.

(b) Describe the motion of the car during the first 9 s.

(c) Suggest reasons why the car reached a steady speed.

(d) Use the graph to estimate how far the car travelled

   (i) in the first 5 s;

   (ii) in the first 9 s.

9 Computing the next move

As predicted, after a journey of several years, a space craft arrives at a planet and sends back pictures. Your holiday aircraft approaches its destination: 'ten minutes to landing' the captain announces. We will tell the story of how movements are computed ahead of time and show you:

- how effects of relative movement can be anticipated

- how to calculate speeds, distances and times when things accelerate

- how to compute the path of an object acted on by gravity

- how to calculate changes of energy of moving objects

The bat catches a moth using ultrasound ranging to keep track of where it is. Somehow the bat has to work out what to do next.

A modern aircraft such as this McDonald Douglas MD11 simulator provides the pilot with a wealth of information but its flight controls may need to be so finely tuned that the human brain and body are simply too slow to cope.

9.1 What's the next move?

You exist in a world of movement—and you live on a moving world. At the end of the 20th century more people are moving faster and further than ever before. Or perhaps not so fast—the average speed of motor traffic in central London is almost exactly what it was for horse-drawn vehicles in 1900. On the other hand, people pop across to New York for the weekend. Over a hundred thousand people cross the English Channel every day.

The fastest and most agile military aircraft need such rapid reactions at the controls that human pilots can't fly them, despite having good reflexes and rigorous training. So computers may have to take over the controls, sensing what the aircraft is doing and acting accordingly. Computing the next move is an elegant and powerful part of physics. But not all movement is to do with traffic. Besides walking and running, many people get pleasure from games demanding the skilful control or prediction of the movement of a ball.

A tennis player, constantly keeping her eye on the ball, computes the optimum moment to make contact for the return shot.

How you catch a ball

The game is on—the ball flies through the air towards you. You have to get hand or racquet in the right place to meet the ball at the right time. How is it done? Most skilful players don't know; they 'just do it'. Practice has taught their minds and muscles how to react faster than they can consciously think. If you are lucky you manage to get your hand to be in the same place as the ball at the same time. The catch is made, the game is won.

A little thinking reveals the hidden rule that your brain and body has learned to use. You 'keep your eye on the ball'. That is, you sense the angle of the

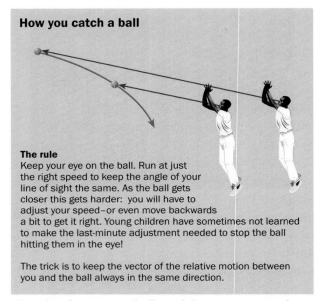

How you catch a ball

The rule
Keep your eye on the ball. Run at just the right speed to keep the angle of your line of sight the same. As the ball gets closer this gets harder: you will have to adjust your speed – or even move backwards a bit to get it right. Young children have sometimes not learned to make the last-minute adjustment needed to stop the ball hitting them in the eye!

The trick is to keep the vector of the relative motion between you and the ball always in the same direction.

direction from eye to ball, and then run so as to keep that angle the same, as the ball moves. That's all. If you can move so as keep the angle to the ball constant then you and it are on a collision course, and you'll catch it (in the eye, if you don't make a last minute adjustment!). The skill comes in rapidly altering how you are running, to keep the ball seeming to come from the same direction. Watch players doing it, or try it yourself, and you'll see how it works. But when you want to win, don't slow yourself down by thinking. 'Just do it'.

Air Traffic Control

Every day, 1200 aircraft enter the air space near London's Heathrow Airport. Each aircraft is identified and its height, distance, direction and speed monitored by radar. Air Traffic Control uses a large and complex computer system to handle all these

data: a prime aim is to predict and act in time to avoid collisions.

You may have seen the radar aerial at an airport turning round and round, surveying the sky for aircraft. Having found an aircraft and measured its distance and velocity, the aerial has to turn right round again before there is another chance to make measurements. So data only come in at intervals. But if the intervals are short it is safe to predict the next position of the aircraft, because the velocity cannot change much in a short time. The controllers can imagine the flight path as a sequence of short straight-line steps.

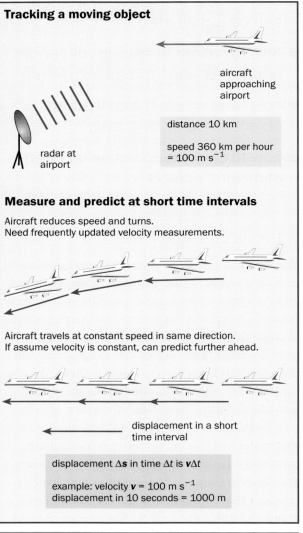

Tracking a moving object

aircraft approaching airport

radar at airport

distance 10 km

speed 360 km per hour
= 100 m s^{-1}

Measure and predict at short time intervals

Aircraft reduces speed and turns.
Need frequently updated velocity measurements.

Aircraft travels at constant speed in same direction.
If assume velocity is constant, can predict further ahead.

displacement in a short time interval

displacement Δs in time Δt is $v\Delta t$

example: velocity v = 100 m s^{-1}
displacement in 10 seconds = 1000 m

● Reconstructing a flight

At velocity v, a step would be a displacement of $v \Delta t$ in time interval Δt. This may be further than you think. An aircraft coming in at 360 km per hour, much less than its typical mid-flight speed, still travels at 100 m s^{-1}. That's one kilometre every ten seconds. Typically the radar dish rotates once every few seconds, say about five seconds.

If Air Traffic Control knows that the aircraft is not going to change speed or turn, the future position can be computed much further ahead. The path is a straight line, moving by equal displacements in equal time intervals. They won't dare predict very far ahead like this, though. It involves the risky belief that the future will be like the past. If the aircraft is changing speed or turning, computing the next move is more work. If the velocity (speed and direction) can be measured at each time interval, Air Traffic Control can look ahead in time just by that time interval. Of course, if they have a flight plan, telling the expected speeds and directions at frequent intervals, they can predict further ahead, plotting a chain of displacement vectors, one for each time interval. But that involves the risky assumption that things will go according to plan. And remember that the displacements could be as much as a kilometre long.

Warning! Collision course

Air Traffic Control has to anticipate possible air collisions, in time to tell the pilots to change direction. And time may be short. Modern planes travel at around 1000 km per hour. Two such planes travelling directly towards one another are coming together at a relative velocity of 2000 km per hour. When 10 km apart there is less than 20 seconds in which to act. That is why computers are needed to look into the future for each aircraft to detect danger as quickly as possible. In fact, Air Traffic Control rules force aircraft on opposing paths to change course before they are about 60 km apart.

Generally, aircraft will approach each other at an angle and with different speeds. Then the question is, *will they collide?* Predicting a collision is predicting something that you want to prevent. So you look ahead, asking a 'what if' question: what will happen if the aircraft go on flying at their present speeds and directions? This requires a calculation which looks into the future.

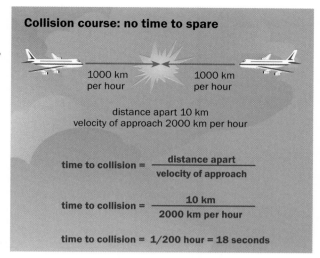

Collision course: no time to spare

1000 km per hour   1000 km per hour

distance apart 10 km
velocity of approach 2000 km per hour

$$\text{time to collision} = \frac{\text{distance apart}}{\text{velocity of approach}}$$

$$\text{time to collision} = \frac{10 \text{ km}}{2000 \text{ km per hour}}$$

$$\text{time to collision} = 1/200 \text{ hour} = 18 \text{ seconds}$$

How is it done? One way is to plot out the two paths step by step and see if they arrive somewhere at the same moment. Another is to use the idea of relative motion.

Relativity at low speeds

You are a passenger in an aircraft. You look out of the window. The Sun glints off the wings of another aircraft far away to the left. A few moments later, the other plane is closer, and you get out your camera to take a photo of it. By the time you have got it in focus it is much nearer and you feel a twinge of alarm. Then you start to get really worried as the other aircraft gets larger and larger in your viewfinder, but—just in time—it zooms ahead and passes in front of your plane.

From your point of view the other aircraft is coming towards you at a fixed angle and speed. Sitting in your plane, you see things from that point of view, as if you were not moving. The movement that you see of the other plane is in fact the relative velocity of the two aircraft. Passengers in the other aircraft would see yours coming towards them.

Both aircraft are moving relative to the ground. In the vector diagram, your plane has been 'stopped' by adding an equal but opposite velocity to its velocity. To put it another way, its velocity has been subtracted from itself to give zero. You can't really 'stop' the plane, of course. You must subtract the same velocity from everything else, including the ground. So with your plane 'stopped', the ground is now slipping away behind you at that velocity. But that's exactly what you seem to see as you look out of the window, thinking of yourself as not moving.

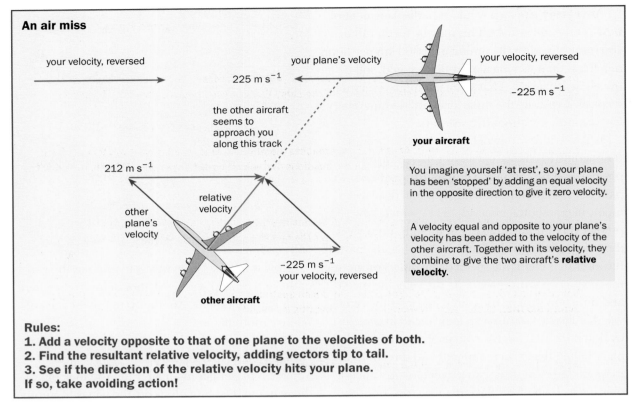

An air miss

your velocity, reversed

your plane's velocity

225 m s^{-1}

your velocity, reversed

–225 m s^{-1}

the other aircraft seems to approach you along this track

your aircraft

212 m s^{-1}

relative velocity

other plane's velocity

–225 m s^{-1}
your velocity, reversed

other aircraft

You imagine yourself 'at rest', so your plane has been 'stopped' by adding an equal velocity in the opposite direction to give it zero velocity.

A velocity equal and opposite to your plane's velocity has been added to the velocity of the other aircraft. Together with its velocity, they combine to give the two aircraft's **relative velocity**.

Rules:
1. **Add a velocity opposite to that of one plane to the velocities of both.**
2. **Find the resultant relative velocity, adding vectors tip to tail.**
3. **See if the direction of the relative velocity hits your plane.**
If so, take avoiding action!

What about the other aircraft? Is it now coming directly towards you? To find out, you subtract your plane's velocity from its velocity too. That is, you find the vector sum of its velocity relative to the ground and the reverse of your velocity relative to the ground. Since your plane has been 'stopped', if that velocity points directly towards you, there is going to be a collision unless something is done about it. The resultant vector shows the **relative velocity** of the other aircraft with respect to yours.

Ships' captains know that there is danger of a collision if the compass bearing of another ship as seen from their ship stays the same. This means that their relative velocity points along the line joining them. It's the same problem as that of catching a ball, except that you are actually trying to get on a collision course with the ball. You run towards the falling ball at just the right velocity to keep its direction from you constant. You—and airline pilots and sailors—need to 'keep your eye on the ball'.

Bats and astronomers

Like airport radar, a bat catching an insect (page 195) is also measuring the distance of a remote object, but by sound ranging rather than radio ranging. Astronomers now measure the size of the solar system by radar. And they are interested in the relative velocities of astronomical objects too—of meteorites which might hit the Earth and of distant galaxies.

In these and other examples, distance is measured not with rulers but with clocks. Time—the travel time of a signal—becomes the measure of distance. This is why astronomical distances are often stated in light-years. And because the speed of light is a fundamental constant, it gives a way to make a fundamental link between space and time. Also, it means that distant objects are seen as they were at earlier times: looking out into space is looking back in time.

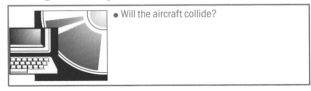

● Will the aircraft collide?

Try these

1 Your car is going at 25 m s^{-1} along a straight narrow road. Another car comes the other way, going at 35 m s^{-1} on its speedometer. What is the relative velocity of the cars? What do you conclude about road safety?

2 Your car heads north at 25 m s^{-1}. Another car approaches along a side road, coming west at 20 m s^{-1}. Both are 100 m from where the roads join. Sketch the positions and velocities of the two cars. What is the velocity of the other car as seen from your car?

3 A canoeist points her canoe directly across a 100 m wide river. She paddles at 2 m s^{-1}; the river flows at 1.5 m s^{-1}. Sketch the velocity vectors. How long does the crossing take? How far downstream does the canoe go in this time?

4 Next time, she does a 'ferry glide'. She points the canoe upstream so that the resultant motion is directly across the river. Sketch velocity vectors to show that this is possible.

5 What is the resultant velocity of the canoe using the ferry glide? (Use a scale diagram, or calculate it.)

6 How long does the canoeist take to cross the river using the ferry glide?

7 Air traffic control detects an aircraft on a bearing 090° (due E). Radar pulses return with a delay of 0.2 m s. 100 s later it is detected on a bearing of 120°, with pulse delay 0.173 m s. What is the displacement vector of the aircraft?

Answers **1.** 60 m s^{-1} **2.** 32 m s^{-1}, at an angle of 51° to the main road **3.** 50 s, 75 m **5.** 1.3 m s^{-1} **6.** 76 s
7. 150 m s^{-1}, bearing 210°

YOU HAVE LEARNED

- That two objects are on a collision course when their relative velocity points along the line joining them.

- How to find the relative velocity of two objects, by subtracting the velocity of one of them from both.

- That velocities add as vectors, tip to tail.

velocity, relative velocity, vector additon

9.2 Speeding up, slowing down

A sprinter obviously wants get up to top speed as soon as possible after the start. Acceleration is the rate of change of velocity. Many accelerations and decelerations go unnoticed around you. For example, as you read this page, your eyes make tiny jumps across the words. To do that, muscles must set your eyes rotating and then stop them again. If you move a hand you have to give it a velocity. If you hit a ball high in the air it slows down going up and then speeds up again coming down.

Most of these accelerations are over quickly. But to think about acceleration it is useful to think about what happens if it continues at a constant rate; if the same amount of velocity is added in the same interval of time. This is the case of **uniform acceleration**. To keep things simple, it's best to begin with acceleration in a straight line.

If you know that the acceleration is constant, you can predict ahead of time where something will get to and how fast it will be moving. Or, as in the case of preventing an air collision (page 198), you can make a 'what if' prediction: what would happen if the acceleration were constant.

A simple way to predict is to draw arrows for the velocity and displacement vectors. Divide time into small slices, and in each small slice of time, add a little to the velocity vector. If the acceleration is uniform,

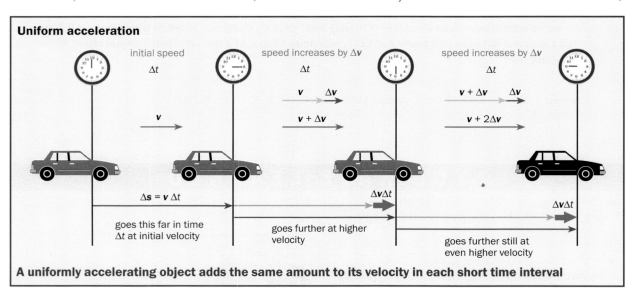

Uniform acceleration

A uniformly accelerating object adds the same amount to its velocity in each short time interval

● Introducing acceleration

the amount added is the same each time. Each new velocity vector is the one from the interval before, with a small fixed extra length of arrow attached, as the object goes faster and faster. The displacement in each time interval is then another vector equal to the velocity at that time multiplied by the time interval.

Reconstructing a traffic incident

An impatient driver is waiting at the lights. They go green, and he accelerates away. Police video and radar show that 12 s later the car was going at 60 miles per hour, past a shop later measured to be 160 m from the lights. The driver claims that the shop is too near the lights for that speed to be possible. Who's right? The incident needs to be reconstructed.

First check: the claimed acceleration is believable. Many cars can accelerate faster than 0–60 mph in 12 seconds. The acceleration was at the rate of 5 mph in each second, on average.

For a different check, convert the acceleration into metric (SI) units. 60 mph is 26.8 metres per second (a mile is 1.609 km). Over the 12 s, the car had to gain speed at a rate of $26.8 \text{ m s}^{-1} / 12 \text{ s} = 2.23 \text{ m s}^{-1}$ per second, on average. The clumsy unit 'metres per second, per second' is abbreviated to m s^{-2}. This is the SI unit of acceleration. 2.23 m s^{-2} is not an impossible acceleration for a car. It is about one fifth of the

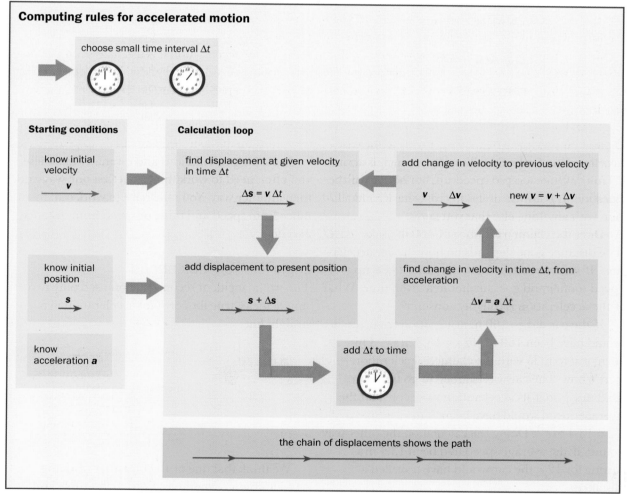

Computing rules for accelerated motion

choose small time interval Δt

Starting conditions

know initial velocity
v

know initial position
s

know acceleration a

Calculation loop

find displacement at given velocity in time Δt
$\Delta s = v \Delta t$

add displacement to present position
$s + \Delta s$

add Δt to time

add change in velocity to previous velocity
v Δv new $v = v + \Delta v$

find change in velocity in time Δt, from acceleration
$\Delta v = a \Delta t$

the chain of displacements shows the path

● Constant acceleration with graphs

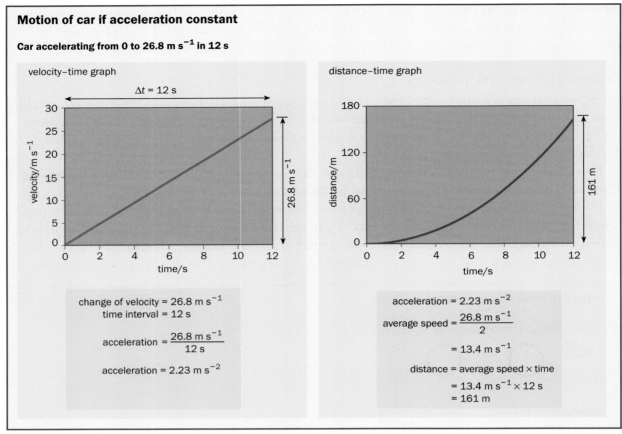

Motion of car if acceleration constant

Car accelerating from 0 to 26.8 m s⁻¹ in 12 s

velocity–time graph

$\Delta t = 12$ s

change of velocity = 26.8 m s⁻¹
time interval = 12 s

$$\text{acceleration} = \frac{26.8 \text{ m s}^{-1}}{12 \text{ s}}$$

acceleration = 2.23 m s⁻²

distance–time graph

acceleration = 2.23 m s⁻²

$$\text{average speed} = \frac{26.8 \text{ m s}^{-1}}{2}$$

$= 13.4$ m s⁻¹

distance = average speed × time

$= 13.4$ m s⁻¹ × 12 s

$= 161$ m

acceleration of a freely falling body, which is about 10 m s⁻². Quite a rapid speed-up, but not out of the question. It is always useful to cross check a calculation with anything else that you know.

Does the claimed distance of 160 m make sense? The distance gone depends on how the car accelerates. If it accelerates a lot at the start, it gets up speed sooner and goes further in a given time. What if the acceleration is roughly constant?

If the car did go 160 m in 12 s its average speed would have been 160 m/12 s = 13.3 m s⁻¹. This turns out to fit in with the claimed acceleration—here's how. The car was claimed to go from 0 to 26.8 m s⁻¹, so if its acceleration were constant the average speed would have been (0 m s⁻¹ + 26.8 m s⁻¹)/2 = 13.4 m s⁻¹. Doing the calculation the other way round, if the average speed had been 13.4 m s⁻¹, lasting for 12 s, the car would have travelled a distance:

$s = vt$
$s = 13.4 \text{ m s}^{-1} \times 12 \text{ s} = 161 \text{ m}$

The reconstruction all seems to fit together. The various measurements and estimates agree. Check them through for yourself to see how they work.

In making calculations and estimates like these, you often need to work things out first one way and then another way. You may have to work out distance from speed and time, or speed from distance and time.

Logic lets you look ahead

The traffic incident reconstruction used only two simple ideas, true for constant acceleration in a straight line:

$$\text{acceleration} = \frac{\text{change of velocity}}{\text{time}}$$
$$= \frac{\text{final velocity} - \text{initial velocity}}{\text{time}}$$
$$\text{average velocity} = \frac{\text{initial velocity} + \text{final velocity}}{2}$$

We think that one of the pleasures of doing physics is how, sometimes, from such simple and almost obvious starting points, you can work out a whole lot of other things. These two simple ideas have several logical consequences, which you wouldn't guess just by looking at them. But with a little work, a lot of consequences can be proved.

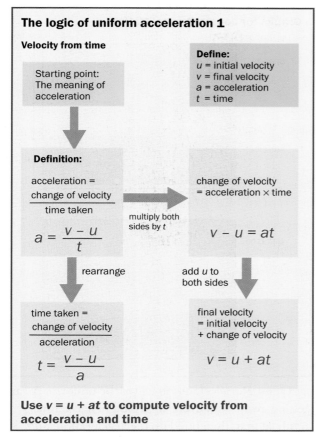

The logic of uniform acceleration 1

Velocity from time

Starting point:
The meaning of
acceleration

Define:
u = initial velocity
v = final velocity
a = acceleration
t = time

Definition:

acceleration =
$\dfrac{\text{change of velocity}}{\text{time taken}}$

$a = \dfrac{v - u}{t}$

multiply both
sides by t

change of velocity
= acceleration × time

$v - u = at$

rearrange

add u to
both sides

time taken =
$\dfrac{\text{change of velocity}}{\text{acceleration}}$

$t = \dfrac{v - u}{a}$

final velocity
= initial velocity
+ change of velocity

$v = u + at$

**Use $v = u + at$ to compute velocity from
acceleration and time**

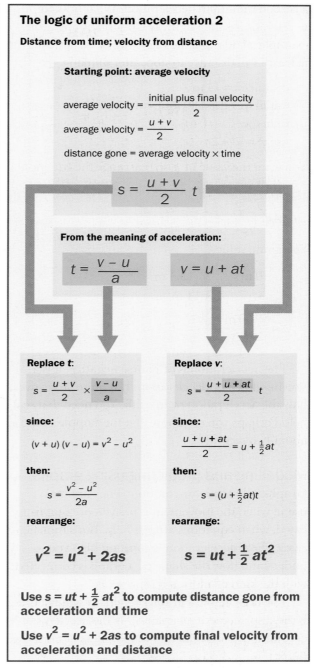

The logic of uniform acceleration 2

Distance from time; velocity from distance

Starting point: average velocity

average velocity = $\dfrac{\text{initial plus final velocity}}{2}$

average velocity = $\dfrac{u + v}{2}$

distance gone = average velocity × time

$$s = \dfrac{u + v}{2}\, t$$

From the meaning of acceleration:

$$t = \dfrac{v - u}{a} \qquad v = u + at$$

Replace t:

$s = \dfrac{u + v}{2} \times \dfrac{v - u}{a}$

since:

$(v + u)(v - u) = v^2 - u^2$

then:

$s = \dfrac{v^2 - u^2}{2a}$

rearrange:

$$v^2 = u^2 + 2as$$

Replace v:

$s = \dfrac{u + u + at}{2}\, t$

since:

$\dfrac{u + u + at}{2} = u + \tfrac{1}{2}at$

then:

$s = (u + \tfrac{1}{2}at)t$

rearrange:

$$s = ut + \tfrac{1}{2}at^2$$

**Use $s = ut + \tfrac{1}{2}at^2$ to compute distance gone from
acceleration and time**

**Use $v^2 = u^2 + 2as$ to compute final velocity from
acceleration and distance**

The consequences are hidden inside these two simple equations. Having brought them out in the open, they are ready to hand to make all sorts of practical predictions about motion.

Useful kinematic equations

The four most useful relationships, all proved just from the meaning of acceleration and the idea of average speed, are:

$$v = u + at$$
$$s = \dfrac{u + v}{2}\, t$$
$$s = ut + \tfrac{1}{2}at^2$$
$$v^2 = u^2 + 2as$$

We are not using the bold symbol notation for vectors here, because it would need extra notation for multiplying vectors. We will continue to use it when vectors are added, especially in two or more dimensions.

Here's a thought. There is no point at all in testing these relationships experimentally. They *can't* be wrong. If the acceleration is constant, they follow logically just from the meanings of acceleration,

speed and distance gone (displacement). So what if you carefully did an experiment and found results that didn't fit them? Could it happen? It could, but it would mean not that the equations are wrong but that the acceleration must not have been constant. That assumption was built into the logic.

Here's an example of putting them to use. The Highway Code suggests safe braking distances for cars going at different speeds, plus of course some 'thinking distance' (see page 216). To be safe, all cars must be designed to be able to achieve at least a certain minimum deceleration. What braking deceleration is the Highway Code assuming? For example, at 30 mph, the Highway Code indicates at least 14 m distance to stop, after thinking.

distance $s = 14$ m
initial speed $u = 30$ mph $= 13.4$ m s^{-1}
final speed $v = 0$ m s^{-1}
use $v^2 = u^2 + 2as$

You want the value of a, so rearrange the equation:

$$2as = v^2 - u^2$$
$$a = \frac{v^2 - u^2}{2s}$$

Substitute values:

$$a = \frac{(0 \text{ m s}^{-1})^2 - (13.4 \text{ m s}^{-1})^2}{2 \times 14 \text{ m}}$$
$$a = \frac{-180 \text{ m}^2 \text{ s}^{-2}}{2 \times 14 \text{ m}} = -6.4 \text{ m s}^{-2}$$

The acceleration is of course negative: it is a slowing down or deceleration. It is quite large: over half the acceleration of free fall. You need that seat belt. Perhaps you can show that the stopping time is about two seconds.

Modelling and predicting using graphs

Graphs help to picture motion. They are most useful when the movement is uneven or complicated, when equations are less help. But simple cases where acceleration is constant help you to understand how the shape of a graph is connected with the kind of motion it represents.

The area below a graph of velocity against time is the displacement (chapter 8). If the acceleration is constant, this area can be thought of as being in two parts:

- a rectangle of height u, width t, area ut which is the displacement there would have been in time t at the initial velocity u
- a triangle, height at, width t, area $\frac{1}{2}at^2$, which is the extra displacement because the velocity is increasing.

● Kinematic equations in vector form

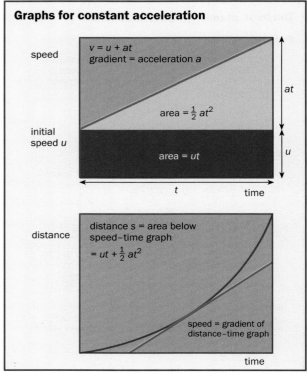

Graphs for constant acceleration

speed

$v = u + at$
gradient = acceleration a

area $= \frac{1}{2}at^2$

initial speed u

area $= ut$

at

u

t time

distance

distance s = area below speed–time graph
$= ut + \frac{1}{2}at^2$

speed = gradient of distance–time graph

time

Often, the acceleration will *not* be uniform. A car getting away from traffic lights might accelerate gently at first, then build up the acceleration, and then finally reduce it as the car gets to a steady speed.

Whatever the shape of the graph

- the *slope* at any point on a distance–time graph gives the *speed* at that point in time
- the *area* under a speed–time graph gives the *distance* travelled.

Graphical methods are useful if you just have observations of distance and time. The speed at any moment can be found by drawing a straight line that has the same slope as the curved distance–time graph at that moment. This is called the *tangent* to the curve and needs a good eye to get an accurate result. The speed v at this time is worked out by calculating the slope $\Delta s / \Delta t$ of the tangent. See chapter 8, page 179.

Free fall

Imagine making a sky-dive. First, you have to pluck up the courage to jump out of the aircraft. You fall faster and faster, accelerating downwards. Soon your downward speed is so great, over 100 miles an hour, that you feel a gale of wind seeming to rush upwards past you. The wind tugs at your clothes.

Graph of a more realistic motion

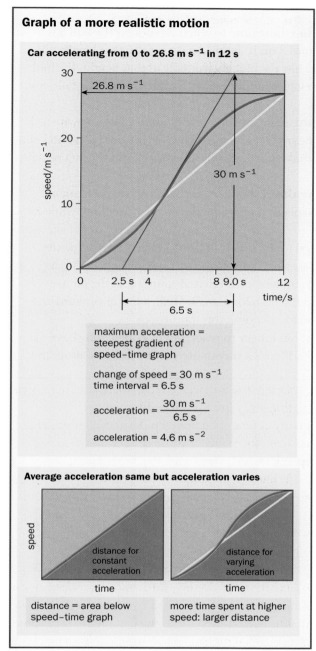

Car accelerating from 0 to 26.8 m s⁻¹ in 12 s

maximum acceleration =
steepest gradient of
speed–time graph

change of speed = 30 m s⁻¹
time interval = 6.5 s

$$\text{acceleration} = \frac{30 \text{ m s}^{-1}}{6.5 \text{ s}}$$

acceleration = 4.6 m s⁻²

Average acceleration same but acceleration varies

distance for
constant
acceleration

distance for
varying
acceleration

distance = area below
speed–time graph

more time spent at higher
speed: larger distance

This parachutist has jumped out of an aircraft at a great enough height for him to enjoy the experience. However, he is **not** in free fall—he has reached a speed where the upward drag of the air on him is equal to the downward pull of gravity, so he goes on falling at this speed.

ground at the same time. There is no atmosphere on the Moon—so no air resistance. Both hammer and feather were able to fall freely. As they did so they went faster and faster. They accelerated at a rate called the *acceleration of free fall*. This is sometimes called the *acceleration due to gravity*.

On Earth the effect of air resistance on a small dense object can be neglected when it falls through small distances. It gets close to being in free fall, accelerating at close to the local maximum rate, the acceleration of free fall (symbol g). On Earth g is close to 9.8 m s⁻². On the Moon g is 1.6 m s⁻².

Try dropping a coin. Watch it carefully. Can you *see* it accelerating? You probably can't, because the time of fall is very short, and you don't have time to 'register' the different speeds. But it *is* accelerating, in free fall. From higher up, the time of fall is longer—say from a bridge overlooking a river. You could drop a small stone and use a stopwatch to time how long it takes to hit the water. Suppose the time of fall is 1.5 seconds, approximately. The stone falls from zero speed and has an acceleration of 9.8 m s⁻². To find the distance, use the relationship:

$$s = ut + \tfrac{1}{2}at^2$$

with $u = 0$, $t = 1.5$ s and $a = g = 9.8$ m s⁻². s is the distance down to the water. Thus:

$$s = \tfrac{1}{2} \times 9.8 \times 1.5^2 = 11 \text{ m}$$

So you know roughly how high the bridge is.

Of course you can't measure the time to as little as 0.1 of a second, so you know only that the water is around 10 m below the bridge. Even a drop of that height takes only between one and two seconds.

Soon its upward drag becomes as big as your whole weight, at which time you no longer fall faster and faster. Your weight and the upward drag balance out and you fall at a steady high speed. 120 miles an hour, or a bit more than 50 m s⁻¹, is typical. To land safely, you will have to open your parachute. With it open, you fall at a much slower steady speed.

If the air weren't there, parachuting would be impossible. Falling objects would just go on falling faster and faster. You may have seen a video of an American astronaut dropping a hammer and a feather on the Moon's surface. Both reached the

Ball games, projectiles and parabolas

Waiting to receive a serve at tennis, you know you will have little time to react. As soon as the ball has been hit, it is falling freely. A serve is hit from a height of around 3 m. If simply let drop from that height, the ball would reach the ground in about 0.8 s. If the server hits the ball horizontally at that height, the ball still takes just 0.8 s to reach the ground. Its horizontal velocity simply carries it towards you as it falls. So you have only 0.8 s or less in which to react.

Actually, the server usually hits the ball at a downward angle, starting it with quite a large downward component of velocity. So the ball reaches the ground near you even sooner, just as it would if it had been thrown downwards. Air resistance and spin of the ball make some further differences, but the fact remains that tennis is a fast-reaction game.

The curve followed by the ball is a mathematical **parabola**. But a tennis ball is no mathematician. It goes along the curve because its free-fall acceleration speeds it up in the downward direction, while its horizontal velocity carries it sideways.

The distance the ball goes horizontally is proportional to the time t. The distance it goes vertically is proportional to t^2. So the shape of the path is the same as that of a graph of t^2 against t, which is a parabola. Or, rather, it has this shape if the forces due to moving through and spinning in the air are unimportant. So a parabola is a good approximation to the path for a cricket ball, a fair approximation for a football or tennis ball and a very bad approximation for a badminton shuttlecock.

If you lob the tennis ball up in the air, your opponent has longer to react, but of course may have to run to get into place to meet the ball coming down again. Here's a way to use vectors to think about the

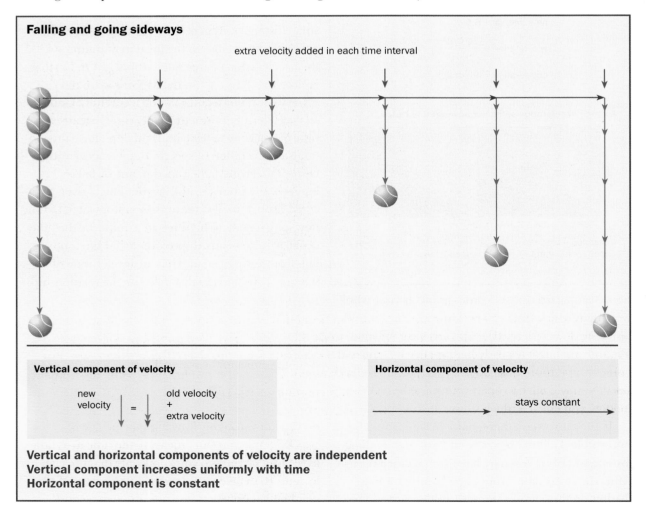

Falling and going sideways

extra velocity added in each time interval

Vertical component of velocity

new velocity ↓ = ↓ old velocity + extra velocity

Horizontal component of velocity

→ stays constant →

Vertical and horizontal components of velocity are independent
Vertical component increases uniformly with time
Horizontal component is constant

path the ball takes. You send the ball up at an angle. If it weren't for the downward acceleration of gravity, it would go on and on upwards at that angle. But in a short interval of time, the downward acceleration will have added a vertical downward component to the velocity. The resultant velocity is the vector sum of the two. So the ball's motion is tilted down a little, and it slows down a little. In the next moment, the same happens, and again, and again, so that the ball's path always curves downwards. Sooner or later, the downward acceleration will have removed all the upward component of velocity, and the ball begins travelling downwards.

The ball keeps on doing two things:
- it travels horizontally with the same horizontal component of velocity
- it accelerates downwards (yes, while going upwards too) with the same acceleration

  That is:
- vectors perpendicular to each other don't affect one another
- only the vertical component of movement is affected by the acceleration of free fall

Tracing out such a path is easy to do with a computer drawing package, just copying arrows and putting them together tip to tail. Draw an arrow to represent the initial displacement at a certain velocity, in a short time interval. Continue the same arrow on from the tip of the first, to see where the ball would get if there were no gravity. But there is gravity. So add a fixed downward displacement, representing the effect of gravity, to the tip of the second arrow. Now draw in the resultant of these two. This is the predicted displacement in the next time interval. Repeat the process, starting each time from the present displacement. Step by step, a parabola appears.

The tennis player intercepts a ball travelling along a parabolic path.

In this fountain in a shopping mall the water streams move in parabolic paths.

This is one way in which accelerated motion can be modelled in a computer. The essential point is breaking down the motion into short straight-line steps. The method is called using finite differences. Time is cut into short intervals and in each interval the velocity is imagined as effectively constant. But going from one interval to the next, the velocity is changed. The change in velocity is calculated from the acceleration at that moment. Although used here for constant acceleration, you can probably see that the method still works if the acceleration varies, as long as you know what it is at each moment.

- Modelling parabolic motion
- Exploring the range of a projectile

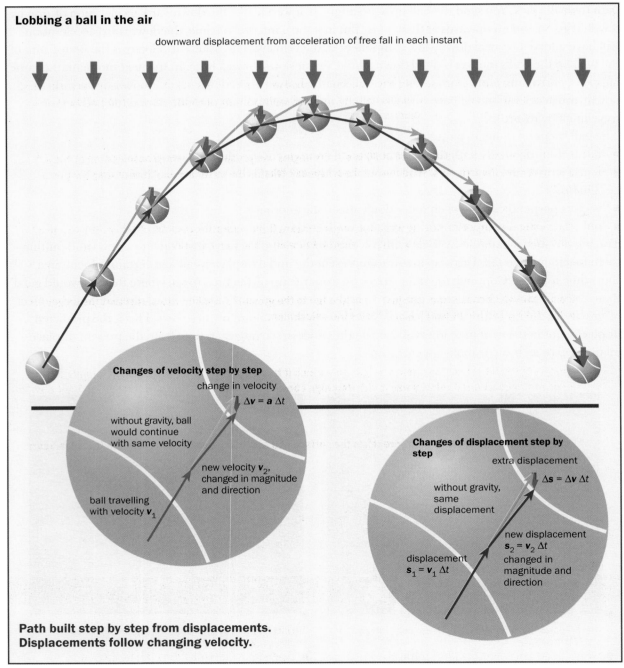

Lobbing a ball in the air

downward displacement from acceleration of free fall in each instant

Changes of velocity step by step

change in velocity

$\Delta \boldsymbol{v} = \boldsymbol{a}\,\Delta t$

without gravity, ball would continue with same velocity

new velocity $\boldsymbol{v}_2$, changed in magnitude and direction

ball travelling with velocity $\boldsymbol{v}_1$

Changes of displacement step by step

extra displacement

$\Delta \boldsymbol{s} = \Delta \boldsymbol{v}\,\Delta t$

without gravity, same displacement

new displacement $\boldsymbol{s}_2 = \boldsymbol{v}_2\,\Delta t$ changed in magnitude and direction

displacement $\boldsymbol{s}_1 = \boldsymbol{v}_1\,\Delta t$

Path built step by step from displacements.
Displacements follow changing velocity.

You may think the step by step computation a bit peculiar. Strictly speaking, it works by imagining switching gravity on and off, which can't be done. As a result, the path computed is kinked, not smooth. In reality, the velocity is changing smoothly all the time, and the path is a smooth curve. By making the time intervals smaller and smaller, you can get closer and closer to the true curve. That isn't quite all. If the velocity is changing all the time, it isn't quite obvious just which change of velocity to calculate and at what moment to add it in. Computer calculations of paths can go quite badly wrong, for this kind of reason.

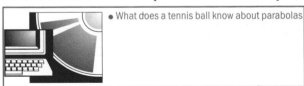

• What does a tennis ball know about parabolas

Try these

1 In a sprint start on a level track a cyclist reaches a speed of 12 m s^{-1} in 3 s. What is the average acceleration? Estimate the distance travelled in this time. Sketch and explain the shape of a possible graph of the cyclist's speed against time, if his maximum speed is 20 m s^{-1}.

2 A light aircraft has a take-off speed of 60 m s^{-1}. Its engine can produce an average acceleration of 4 m s^{-2}. How long does the aircraft take to reach take-off speed? What is the minimum length of runway the plane needs?

3 Write a few sentences discussing what difference it makes if the aircraft in question 2 takes off into a head wind blowing at a speed of 10 m s^{-1}. What about a tail wind of the same speed?

4 From what height does a ball take just 1 s in free fall to the ground? A student calculates that the height for a drop taking 100 s is about 50 km. Criticise the calculation.

5 A tennis ball is hit horizontally from a height from which it takes 0.6 s to fall to the ground. If it lands 10 m away, at what horizontal velocity was it hit? How high above the ground was it halfway through its flight time?

6 Estimate the speed at which a diver reaches the surface of a swimming pool after diving from the 10 m board. Convert to km per hour.

Answers 1. 4 m s^{-2}, 18 m **2.** 15 s, 450 m **4.** 4.9 m **5.** 16.7 m s^{-1}, 1.3 m **6.** 14 m s^{-1}, 50 km per hour

YOU HAVE LEARNED

- How to construct paths of accelerated bodies step by step.

- How to use the kinematic equations to predict displacements and velocities for uniform acceleration.

- That the area under a speed–time graph can give the distance travelled.

- That the slope of a speed–time graph can give the acceleration.

- That the slope of a distance–time graph can give the speed.

- That a projectile in free fall follows a parabolic path.

- That vertical and horizontal components of velocity of a projectile are independent.

speed–time graph, distance–time graph, projectile, free fall

9.3 Force, mass, gravitation

Sitting at the end of the runway, your aircraft's engines go to full throttle. The brakes come off, and the back of your seat pushes you forward as the plane gathers speed for take-off. The holiday has begun. A typical acceleration is 3 m s^{-2}, similar to that of a sports car (0–60 mph in 8.9 s), but the aeroplane's acceleration goes on a good bit longer.

Runway lengths at civil airports don't differ much, and take-off speeds of different civil jet aircraft are much the same as one another, so the big planes need to accelerate as fast as the small ones. For that reason, the more massive planes need more powerful engines, which provide more thrust. For example the huge Boeing 747, with maximum take-off mass 375 000 kg, has engines which provide a thrust of over 1000 kN (the weight of ten 10 tonne lorries). The mass of the much smaller Boeing 737 is about five times less than that of the 747, and its engines in fact produce about one fifth the thrust.

You have met the same thing very often. You can't easily start running quickly carrying a full suitcase, and it's harder still if you are carrying two—both have to be accelerated. You get used to pushing a younger brother or sister on a swing; then one day they have grown bigger and it's harder to do. Kick a big stone or hit a big punch-bag, and you hurt your toe or fist, while it hardly gets moving at all. Trains are very massive, and despite powerful locomotives they accelerate rather slowly out of the station.

Putting this all together means that you need a bigger force to give the same acceleration to a bigger mass, and you need a bigger force to give a bigger acceleration to the same mass. Force **F**, mass m and acceleration **a** are linked by the simple relationship:

$$force = mass \times acceleration \quad \text{or} \quad \bm{F} = m\bm{a}$$

Force and acceleration are vectors. Mass is a scalar, not a vector. The usual units are:

acceleration **a** in metres per second each second (m s^{-2})
mass m in kilograms (kg)
force **F** in newtons (N)

The equation explains why the design thrust of an aeroplane is roughly proportional to its mass. Take-off accelerations are similar, so roughly $\bm{F} \propto m$.

A sprint start

The relationship *force = mass × acceleration* helps analyse how well an athlete performs. The graph shows data for a sprinter accelerating from the starting blocks, taken from a video. For the first

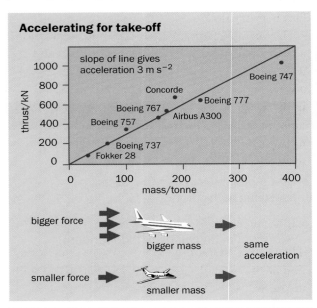

Accelerating for take-off

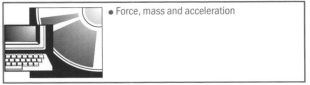

• Force, mass and acceleration

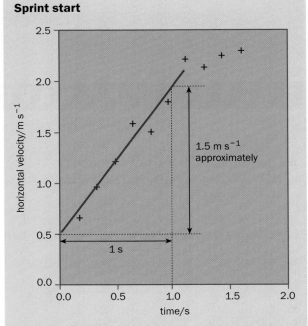

Sprint start

This graph was produced from a video recording of a sprint start. It shows the movement of a point at the runner's waist.

second the athlete's horizontal velocity increases fairly uniformly, increasing by nearly 1.5 m s^{-1}. His acceleration is about 1.5 m s^{-2}. This sprinter looks fairly massive: a guess at his mass might be 80 kg, say. To produce this acceleration would need his leg muscles to produce an average force of 120 N. This isn't all that big a force. It's what you need to pick up a dozen 1 kg bags of sugar.

The sprinter's body weight is 800 N, so the driving force of 120 N is 15% of the force he would need to lift himself, say going upstairs. To accelerate as fast as possible, the sprinter uses starting blocks which help the initial driving force from his legs to be as much as possible in a horizontal direction, and he leans forwards for the first few paces of the race.

Initial acceleration is the key to winning. The best accelerator will be ahead of the field after the first second or so. The world's best athletes will aim to run 100 m in less than 10 seconds. This is an average speed of at least 10 m s^{-1}, so the sprinter must accelerate to a bit more than that speed inside a second or so, and keep it up for the next 9 s. Clearly the secret to reaching Olympic standard in the 100 metres is a good 'power to weight ratio'. The ideal is a light body with powerful leg muscles. Powerful muscles are big muscles. But big muscles

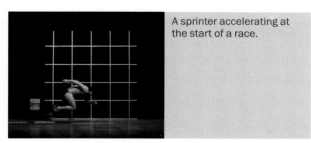

A sprinter accelerating at the start of a race.

are heavy. Top class male sprinters are usually large, tall men. Longer legs mean more efficient use of muscles (using long bones as levers) and less energy loss because the feet don't hit the ground so often. Each stride also covers more ground.

Let's return to the sprinter in the photograph. The end of the graph suggests that the sprinter has reached a maximum speed of about 2.5 m s^{-1}, so perhaps he wasn't trying too hard.

Mass and weight: on Earth, on the Moon

Two quite different things can happen to a lump of matter, both depending on its mass:

- it can be pushed by a force and accelerate: it may be hard to get moving
- it can be pulled down towards the Earth by gravity: it feels heavy.

Try this: imagine getting ready for an aerobics or gymnastics class. Stand still and just let yourself feel your weight in the soles of your feet. Tell yourself that this feeling is the Earth pulling you down. Try to feel the sensation in your feet, not as you pushing down but as a pull coming from the Earth below. That's the action of gravity: your *weight*. Now spring a little from side to side. Feel the pushes and pulls you have to give to set your body moving one way, to stop it and start it going the other way. That's due to your *mass*, needing a force to change its velocity, to accelerate it. Sometimes you'll see this called 'inertial' mass (from a Latin word meaning 'lazy').

Let us tell you what this would feel like if you did it on the Moon, where the gravitational pull is six times smaller. The feeling of weight on your feet would be six times less—you would need much less effort to lift yourself on your toes. But moving your body to and fro would feel *exactly the same* as on Earth! So you could easily lift a fellow astronaut off the ground, but it would be just as hard as usual to pull them quickly towards you. In a Moon race, you could jump higher but you could not do a much faster sprint start.

If you often went to the Moon, you would have a clear idea of the difference between mass and weight. As it is, staying on Earth, the two seem always to go together and are hard to tell apart. And in shops, the 3 kg mass of a bag of potatoes—the amount of potato there is—is always actually called its *weight*, instead. There's a good reason why this is what people do. It is that:

the pull of gravity *mg* on an object is proportional to its mass
the force *ma* needed to accelerate an object is proportional to its mass.

The result is that the weight is always a good indication of the mass, if the gravity field you live in stays the same. One important difference is that:

weight *mg* is a force—the pull of gravity—and has units newtons, N
mass *m* is an amount of matter, and has units kilograms, kg.

The gravity field

The relationship $F = ma$ helps us to understand and predict the motion of cars, aeroplanes, trains and sprinters. But it is interesting to think that the start of it all was with Galileo and Newton. What Newton was trying to understand was movements in the heavens, not getting about on Earth. To explain them, he developed the idea of force being proportional to mass and acceleration. To account for the forces on planets and stars he imagined a force acting in empty space: the force of gravity. The idea is used today to calculate how to place a TV satellite so that millions of satellite dishes can point to it (chapter 11, A2 Student's book).

Try to picture gravity. You know that everywhere there is a downward pull towards the Earth. It's what makes cups fall and break, and doing the high jump difficult. How to picture an invisible something everywhere which pulls things downward? One way is to imagine a downward arrow at each point in space. You can't draw arrows everywhere, so why not draw them equally spaced? Then the equal spacing suggests that the pull is the same everywhere, as does the equal length of the arrows.

This picture of a field filling space is one of the great imaginative leaps made in physics. It led in the end to ideas about electric and magnetic fields, about electromagnetic waves, and to modern theories of the nature of fundamental particles. Yet it seems crazy. How could empty nothingness *do* anything? No wonder Newton's contemporaries, especially the French, were very critical of it.

Objects of different mass fall freely with the same acceleration. To give a 10 kg mass the same acceleration as a 1 kg mass needs 10 times the force. By a neat coincidence the force of gravity on the 10 kg mass is exactly 10 times greater, so that the end result is the same: both masses accelerate downwards at the same rate of $9.8~\text{m s}^{-2}$. A 'neat coincidence'? Not really. It says that mass as measured by accelerations is exactly the same as mass as measured by gravitational pulls.

This means that the strength of the gravity field can be measured by the acceleration of free fall. At least, it can if effects of the rotation of the Earth can be ignored. The units of the gravity field strength are N kg^{-1}. The units of acceleration are m s^{-2}. They look different but are really the same.

Forces change velocities

Newton explained that whenever a velocity is changed a force must be involved. Sometimes the change is simply a change in speed—the moving object gets faster or it slows down. But a velocity can change without a change in speed. This is where the idea that velocity is a vector really bites. Adding a change to a vector can change its direction as well as changing its magnitude: that's vector addition by the tip-to-tail rule.

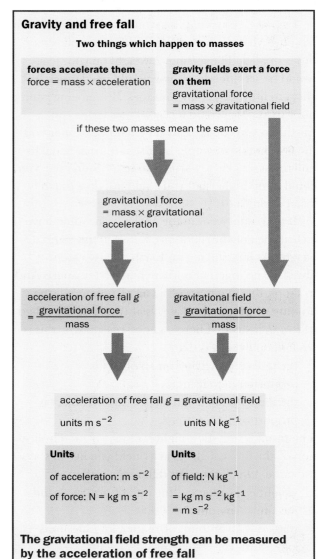

Gravity and free fall

Two things which happen to masses

forces accelerate them
force = mass × acceleration

gravity fields exert a force on them
gravitational force
= mass × gravitational field

if these two masses mean the same

gravitational force
= mass × gravitational acceleration

acceleration of free fall g
$= \dfrac{\text{gravitational force}}{\text{mass}}$

gravitational field
$= \dfrac{\text{gravitational force}}{\text{mass}}$

acceleration of free fall g = gravitational field

units m s^{-2} units N kg^{-1}

Units
of acceleration: m s^{-2}
of force: $\text{N} = \text{kg m s}^{-2}$

Units
of field: N kg^{-1}
$= \text{kg m s}^{-2}\,\text{kg}^{-1}$
$= \text{m s}^{-2}$

The gravitational field strength can be measured by the acceleration of free fall

● Why some things move in circles

Different masses fall with the same acceleration

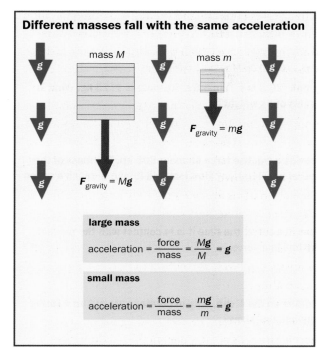

mass M

mass m

$F_{\text{gravity}} = m\boldsymbol{g}$

$F_{\text{gravity}} = M\boldsymbol{g}$

large mass

$$\text{acceleration} = \frac{\text{force}}{\text{mass}} = \frac{M\boldsymbol{g}}{M} = \boldsymbol{g}$$

small mass

$$\text{acceleration} = \frac{\text{force}}{\text{mass}} = \frac{m\boldsymbol{g}}{m} = \boldsymbol{g}$$

So, according to Newton, **forces change velocities**—sometimes in magnitude (speed), sometimes in direction, sometimes both.

This is connected with an idea of Galileo's, which most people find very shocking. No force, no change in velocity. But no *change* in velocity doesn't mean *no velocity*. If no force acts things can just keep moving. So the idea is that *movement itself needs no force*. Just moving isn't special, if only because velocities are all relative.

This is a far cry from the world of horse-driven carts and coaches being pulled with 'great difficulty along the muddy and rutted roads of 17th century Europe. It's an equally far cry from the high speed trains or aeroplanes of today. You never get the chance to experience the complete absence of forces, so it is hard to believe. Your belief might increase for a moment if you found yourself gliding on skates towards the barrier round an ice-rink, and hadn't learned how to dig in the skates to slow down. Maybe really to see the point you have to look to the stars, like Newton. What 'keeps them moving'? Nothing. They just keep going unless a force changes their velocity.

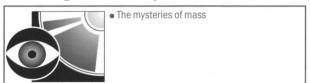

● The mysteries of mass

Change of speed; change of direction

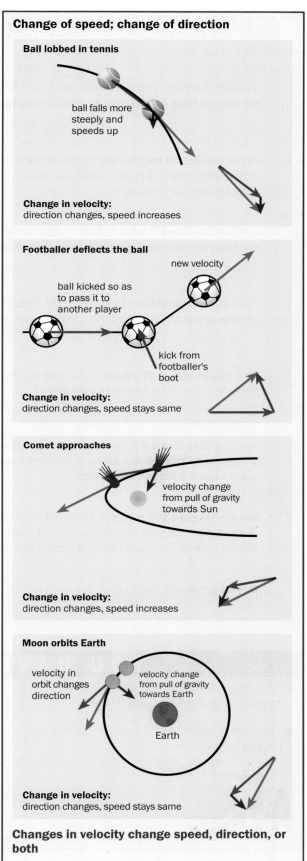

Ball lobbed in tennis

ball falls more steeply and speeds up

Change in velocity:
direction changes, speed increases

Footballer deflects the ball

new velocity

ball kicked so as to pass it to another player

kick from footballer's boot

Change in velocity:
direction changes, speed stays same

Comet approaches

velocity change from pull of gravity towards Sun

Change in velocity:
direction changes, speed increases

Moon orbits Earth

velocity in orbit changes direction

velocity change from pull of gravity towards Earth

Earth

Change in velocity:
direction changes, speed stays same

Changes in velocity change speed, direction, or both

Try these

Gravitational field strength on Earth = 9.8 N kg^{-1}; gravitational field strength on Moon = 1.6 N kg^{-1}.

1 An Aston Martin DB7 coupé accelerates from 0 to 60 mph (26.8 m s^{-1}) in 5.8 s. Its mass is 1725 kg. What is the average force accelerating it? What percentage is this of its weight?

2 The locomotives of the Channel Tunnel shuttle train provide a tractive force of about 800 kN. The mass of the train is 2400 tonne (1 tonne = 1000 kg). What is its initial acceleration? How long will it take to reach a speed of 20 km per hour from rest?

3 Estimate: the mass of a tennis ball; the speed it reaches in a serve; the time it is in contact with the racquet. Use these to estimate the average force exerted on the ball in a serve.

4 What is the acceleration of free fall on the Moon? Standing on the Moon, 5 kg of potatoes are put on a spring balance calibrated in kg on Earth. What does the spring balance show?

5 On Earth, you find that you can easily jump 0.1 m vertically off the floor. What is your initial velocity? How high would the same jump take you on the Moon?

6 You jump off a step 0.5 m above the ground. With what velocity do you hit the ground? You reduce the force of the impact by bending your knees as you land. Over what time must you spread the impact if you want the force not to exceed twice your body-weight?

7 You hope to run a 100 m race in 15 seconds. Estimate the top steady speed at which you must run. You reach this speed 1.5 seconds from the start. What (average) horizontal force must your muscles produce as you accelerate from the start? (Assume your mass is 55 kg).

Answers 1. 8 kN; 46% **2.** 0.33 m s^{-2}; 17 s **4.** 1.6 m s^{-2}, 0.82 kg **5.** 1.4 m s^{-1}, 0.61 m **6.** 3.1 m s^{-1}; 0.16 s **7.** 6.7 m s^{-1}; 244 N

YOU HAVE LEARNED

- That force, mass and acceleration are related by the relationship **F = ma**.

- That the gravitational force on an object is proportional to its mass.

- That different masses have the same acceleration **g** due to gravity.

- That the gravitational field **g** is the gravitational force per unit mass.

- That the gravitational field near the surface of the Earth is uniform.

free fall, mass, weight

9.4 Transport engineering

You are driving down a quiet street. Suddenly a child runs into the road ahead of you. Can you stop in time? Car manufacturers have to design brakes so that a car can stop with better than a minimum deceleration. Using $\boldsymbol{F} = m\boldsymbol{a}$, this means that the greater mass of the car, the greater the braking force must be.

The Highway Code suggests typical stopping distances for cars. The stopping distance is made of two parts. One part is the distance the car goes while the driver is reacting, before the brakes go on. This 'thinking time' is the same at any speed, so the thinking distance is proportional to the speed.

The other part is the distance needed to stop the car with the brakes on. This increases a lot as the speed increases. The brakes have to remove all the kinetic energy from the car, and its kinetic energy is

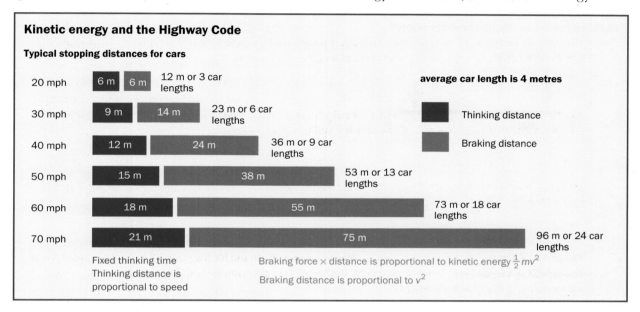

Kinetic energy and the Highway Code

Typical stopping distances for cars

20 mph	6 m / 6 m	12 m or 3 car lengths
30 mph	9 m / 14 m	23 m or 6 car lengths
40 mph	12 m / 24 m	36 m or 9 car lengths
50 mph	15 m / 38 m	53 m or 13 car lengths
60 mph	18 m / 55 m	73 m or 18 car lengths
70 mph	21 m / 75 m	96 m or 24 car lengths

average car length is 4 metres

■ Thinking distance
■ Braking distance

Fixed thinking time
Thinking distance is proportional to speed

Braking force × distance is proportional to kinetic energy $\frac{1}{2}mv^2$

Braking distance is proportional to v^2

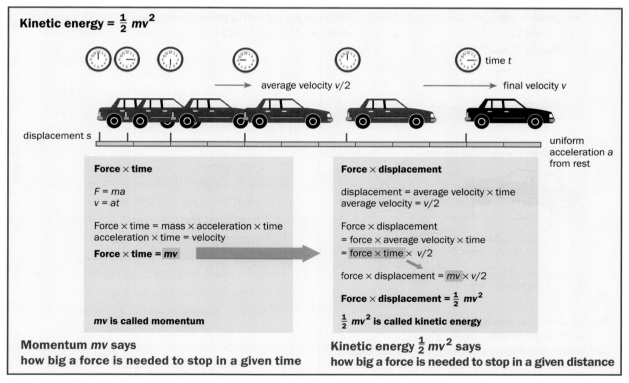

Kinetic energy = $\frac{1}{2}mv^2$

average velocity $v/2$ · · · final velocity v · · · time t

displacement s

uniform acceleration a from rest

Force × time

$F = ma$
$v = at$

Force × time = mass × acceleration × time
acceleration × time = velocity

Force × time = mv

mv is called momentum

Force × displacement

displacement = average velocity × time
average velocity = $v/2$

Force × displacement
= force × average velocity × time
= force × time × $v/2$

force × displacement = mv × $v/2$

Force × displacement = $\frac{1}{2}mv^2$

$\frac{1}{2}mv^2$ is called kinetic energy

**Momentum mv says
how big a force is needed to stop in a given time**

**Kinetic energy $\frac{1}{2}mv^2$ says
how big a force is needed to stop in a given distance**

proportional, not to the speed but to the square of the speed. This explains the very large stopping distances needed for high speeds.

You can see this from the kinematic equation $u^2 = -2as$ (with final velocity v equal to zero) For fixed deceleration, the stopping distance is proportional to u^2. The kinetic energy is the energy the car has when going at velocity v. To calculate how much this is, you need to know how much energy was put in as it speeded up. That is, you need

to know what force accelerated it and over what distance the force acted. Then the energy going from fuel burning with oxygen to the motion of the car is

energy transferred = work done = force × distance

To calculate force × displacement it is easiest to start by calculating force × time instead. Since $F = ma$, and at is the increase in velocity v, then starting from rest

force × time = mv

The force multiplied by the time is an important quantity, called the momentum. It tells you how much time you have to stop the car. But what you need to calculate the kinetic energy is force × *displacement*. This can be found very neatly, by noticing that the displacement is just the average velocity multiplied by the time.

force × displacement =
force × time × average velocity

If the acceleration is uniform, the average velocity is $v/2$, so

force × displacement =
$mv \times (v/2) = \frac{1}{2}mv^2$

This is the **kinetic energy** of a body moving at speed v. The kinetic energy of the Channel Tunnel train, mass 2400 tonnes, speed 140 km per hour, is nearly 2000 MJ. It takes one and a half kilometres to come to a stop with the brakes full on. You should be able to work out the braking force and the deceleration.

Why force × displacement?

If you are trying to stop a car before you hit something, it is obviously the distance in which you have to stop that matters. But there are more basic reasons why force × displacement is used to calculate work done.

Do you know how to lift someone with just one finger? Stand them on a plank, with a fulcrum under it near their feet, and push down with a finger on the other end. You don't have to push hard, so you don't hurt your finger. But there's a price to pay for using the plank as a lever to magnify the force. While you push down quite a long way, the person is lifted by only a small amount. Forgetting any friction, and the weight of the plank, the force × displacement is the same at both ends. The work you put in comes out unchanged at the other end.

Making forces or displacements bigger

Lever: magnifying a force

big force F small force f

small displacement d big displacement D

$f \times D \geq F \times d$
work in ≥ work out

small force f in
big force F out

big displacement D in
small displacement d out

Bicycle: magnifying a movement

$F \times d \geq f \times D$
work in ≥ work out

small distance d moved by foot on pedal
large distance D moved by wheel on road

large force F of foot on pedal
small force f of wheel on road

Car hydraulic jack: magnifying a force

$f \times D \geq F \times d$
work in ≥ work out

jack lever moved through large distance D
car lifted through small distance d

small force f exerted on jack handle
very large force F lifts car

Forces can be made bigger.
Displacements can be made bigger.
But force × displacement cannot be made bigger.

Your bicycle works the other way round. The chain and wheels magnify the distance moved by your feet as you push the pedals round, into a bigger distance as the wheels carry you along. The price is that you have to push harder on the pedals than the bicycle is pushed forward. That's why you can't accelerate very fast on a bike. Again, the energy—force × distance—which you put in cannot be increased.

Designing railway gradients

When the railway system was built in the 19th century, large numbers of cuttings or tunnels through hills, and embankments above valleys, had to be built. Even as you travel by train today, you keep finding yourself looking down or up at the surrounding countryside. This labour was first done with picks and shovels by 'navvies'—the name is short for 'navigators' which was what people called the labourers who previously dug the canal system ('navigations'). All this digging was expensive, so the transport engineers like Isambard Kingdom Brunel planned the routes of their railways very carefully, going along contours of the land as much as they could.

The reason to take all this trouble is that trains are very bad at going uphill. Trains are very heavy, and it takes a lot of energy to drag them uphill. The amount of energy needed is easily calculated, from force × displacement. If the mass of the train is m,

'Navvies' digging out a railway cutting, using only hand tools and muscle power. Today, machines replace large numbers of manual workers.

its weight is mg, and if it goes a vertical distance h as it goes uphill, then:

Force to lift train = mg

Displacement in direction of force (vertical) = h

Work done = force × displacement = mgh

Energy going to gravitational potential energy = mgh

The design of the Channel Tunnel link is no exception. The tunnel has to dive under the Channel and come back up again, as well as getting through or round hills on the way. The maximum permitted gradient for Le Shuttle is 1 in 90. So if the track goes up or down by 100 m that height change must be spread over 9 km. The working mass of Le Shuttle is 2400 tonnes, a weight mg of about

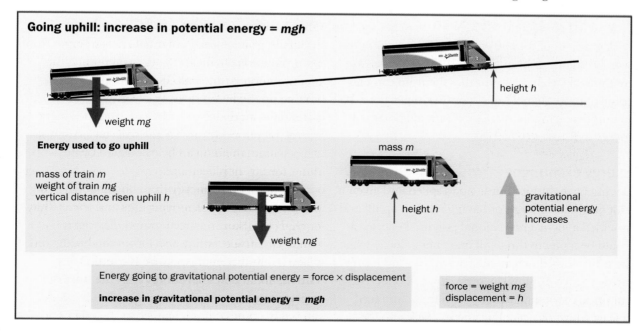

Going uphill: increase in potential energy = mgh

height h

weight mg

Energy used to go uphill

mass of train m
weight of train mg
vertical distance risen uphill h

mass m

weight mg

height h

gravitational potential energy increases

Energy going to gravitational potential energy = force × displacement

increase in gravitational potential energy = mgh

force = weight mg
displacement = h

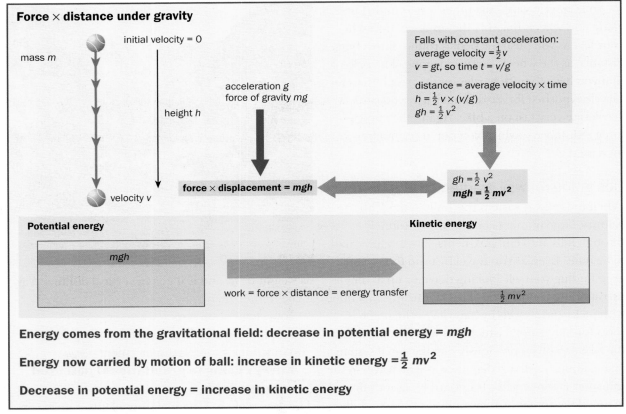

Force × distance under gravity

initial velocity = 0

mass *m*

acceleration *g*
force of gravity *mg*

height *h*

velocity *v*

Falls with constant acceleration:
average velocity = $\frac{1}{2}v$
$v = gt$, so time $t = v/g$

distance = average velocity × time
$h = \frac{1}{2}v \times (v/g)$
$gh = \frac{1}{2}v^2$

force × displacement = *mgh*

$gh = \frac{1}{2}v^2$
$mgh = \frac{1}{2}mv^2$

Potential energy   **Kinetic energy**

mgh

$\frac{1}{2}mv^2$

work = force × distance = energy transfer

Energy comes from the gravitational field: decrease in potential energy = *mgh*

Energy now carried by motion of ball: increase in kinetic energy = $\frac{1}{2}mv^2$

Decrease in potential energy = increase in kinetic energy

24 MN. Its locomotives deliver a maximum of 11.2 MW, or 11.2 MJ per second. Suppose about half its power, say 6 MJ per second, is being used to go uphill, the rest being used to overcome resistance to motion. Suppose the train is going up a slope and is rising vertically by distance *h* each second:

energy provided per second = 6 MJ per second
energy needed per second = 24 MN × *h*
h = 6 MJ per second/24 MN = 0.25 m per second

A gradient of 1 in 90 will spread this 0.25 m rise over a distance of about 22 m. If these guesses are anything like right, it seems that the train can't go uphill at more than 22 m s⁻¹, or 80 km per hour. Its locomotives can't increase gravitational potential energy any faster.

Energy exchanges

Having brought Le Shuttle up to speed and given it a lot of kinetic energy, or having taken it uphill and provided a lot of gravitational potential energy, it would be a pity to throw all that energy away when the train slows down or runs downhill. And it isn't thrown away. To slow the train down, its electric motors are switched to act as dynamos, and feed at least a part of its kinetic energy back into the overhead electrical supply lines. Similarly, going downhill Le Shuttle can provide electrical power back to the National Grid, as the gravitational potential energy decreases. The train to Paris may be helping to keep your lights on, even at this moment.

Energy exchanges between gravitational potential energy and kinetic energy are very common. They happen every time a ball falls under gravity. As it falls, gravitational potential energy comes from the gravitational field and goes to increasing the kinetic energy of the ball. If no energy goes to stirring up the air on the way, the gravitational potential energy decreases by *mgh*, and the kinetic energy increases by $\frac{1}{2}mv^2$ (if the ball started at rest). The changes in both can be calculated by the work done, force × displacement.

You are used to energy from burning fuel being put into the energy of moving cars and trains. But energy can also come from gravity. Energy is needed to lift something away from the Earth, and when it falls that energy can be retrieved. Gravitational potential energy acts as a store of energy.

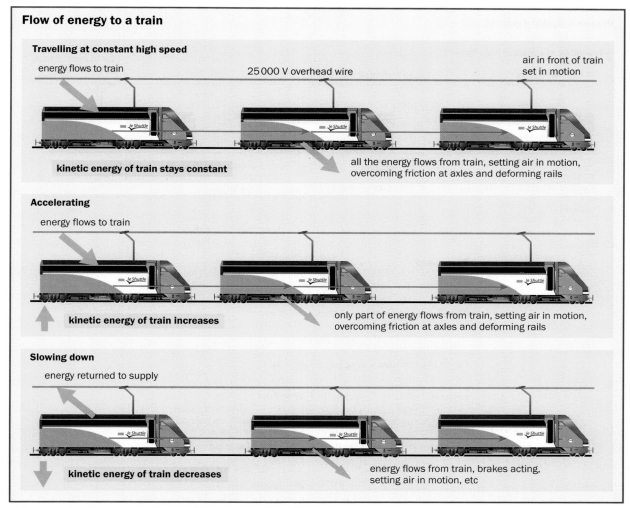

Flow of energy to a train

Travelling at constant high speed

energy flows to train

25 000 V overhead wire

air in front of train set in motion

kinetic energy of train stays constant

all the energy flows from train, setting air in motion, overcoming friction at axles and deforming rails

Accelerating

energy flows to train

kinetic energy of train increases

only part of energy flows from train, setting air in motion, overcoming friction at axles and deforming rails

Slowing down

energy returned to supply

kinetic energy of train decreases

energy flows from train, brakes acting, setting air in motion, etc

A trip to Paris: powering a high speed train

Travelling to Paris one Spring weekend, your mind is full of plans for what you are going to do. Inside the train, you'd hardly know you were moving, even though you see the countryside flashing past out of the corner of an eye. It's all so smooth that you don't think about the large flows of energy to and from the train. But think now about what is happening. As the train pushes into the air in front of it, it sets that air moving, probably at a speed similar to that of the train. A surprising amount of air is set moving. The train can travel least 50 m in a second (180 km per hour). Its frontal area might be 20 m^2. That would be 1000 m^3 of air pushed aside per second, with a mass of 1.3 tonnes (1300 kg). Set moving at 50 m s^{-1} the energy $\frac{1}{2}mv^2$ comes to 1.6 MJ per second, or 1.6 MW.

Add to this the dragging of air by the sides of the train, friction heating the axles and rails, and energy used in deforming the track as the heavy train goes over it, and you soon reach the 10 MW or so of power needed to keep the train running.

When the train is running at a steady speed, energy flows out as fast as it flows in. If the driver increases the input power, more energy comes in than goes out, and the difference increases the kinetic energy of

• Modelling air resistance

• Your toolkit for computing the next move

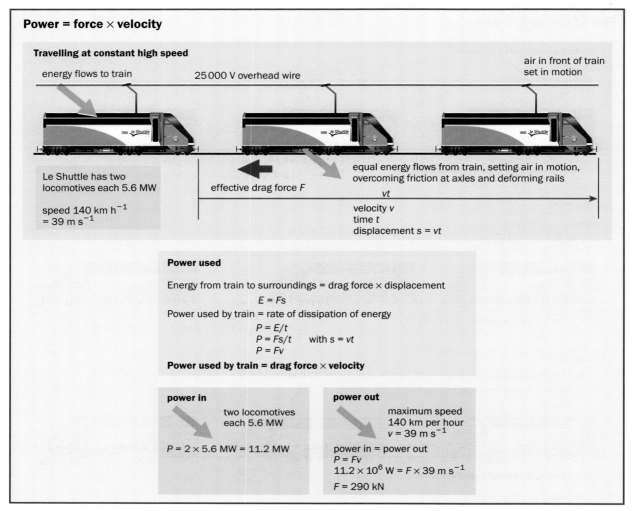

Power = force × velocity

Travelling at constant high speed

energy flows to train   25 000 V overhead wire   air in front of train set in motion

equal energy flows from train, setting air in motion, overcoming friction at axles and deforming rails

effective drag force F

Le Shuttle has two locomotives each 5.6 MW

speed 140 km h^{-1} = 39 m s^{-1}

velocity v
time t
displacement $s = vt$

vt

Power used

Energy from train to surroundings = drag force × displacement
$$E = Fs$$
Power used by train = rate of dissipation of energy
$$P = E/t$$
$$P = Fs/t \quad \text{with } s = vt$$
$$P = Fv$$

Power used by train = drag force × velocity

power in

two locomotives each 5.6 MW

$$P = 2 \times 5.6 \text{ MW} = 11.2 \text{ MW}$$

power out

maximum speed 140 km per hour $v = 39$ m s^{-1}

power in = power out
$$P = Fv$$
$$11.2 \times 10^6 \text{ W} = F \times 39 \text{ m s}^{-1}$$
$$F = 290 \text{ kN}$$

the train. If the driver turns off the input power and puts on the brakes, energy goes rapidly from the train to the surroundings, and its kinetic energy decreases.

The power needed to keep the train running at constant speed is just the power needed to replace all the energy the train is losing each second.

work done = drag force × displacement

work done per second = drag force × distance moved per second

But power = work done per second, and velocity = distance moved per second, so:

power = drag force × velocity or $P = Fv$

The modern Galileo

See if you can imagine yourself as a modern Galileo, arguing that things can go on moving for ever if no force acts on them. The modern Galileo might be a railway engineer. He knows that improving the streamlining of a train reduces the drag on the train and so reduces the tractive force needed to keep it going. Suppose he imagines super-streamlining which reduces the air drag more and more. Also super-lubricants which remove axle friction. The rails are made smoother and smoother, and very rigid. Soon in fantasy, the engineer-Galileo has a train running at hundreds of kilometres an hour off a car battery.

Then the last step. Imagine the drag forces are reduced to zero. Then the train will just go on running, with nothing to stop it, and no need for anything to power it. Impossible? Certainly it's only a dream for trains, but look at the stars. Think of the Earth turning day after day. Motion without forces just goes on and on. Motion at constant velocity needs no cause to keep it going.

Try these

1 Calculate the kinetic energy of (a) a 100 g tennis ball travelling at 30 m s^{-1}, (b) a 50 kg runner doing a 1000 m race in 4 minutes, (c) a 500 kg car travelling at 2 m s^{-1}.

2 Calculate the change in gravitational potential energy when (a) a 50 kg athlete jumps 1.5 m above the ground, (b) a 1000 kg car is jacked 100 mm above the ground, (c) a 200 g ball is hit 20 m up in the air.

3 A car of mass 1235 kg can accelerate at 2.1 m s^{-1}. What is the tractive force acting? If the car accelerates at this rate over a distance of 50 m, how much kinetic energy does it acquire? What is its speed?

4 A train of mass 1000 tonnes starts going uphill at 10 m s^{-1}. The extra power required to keep it going is 1 MW. Estimate the steepness of the slope.

5 If a swimmer generating 200 W of 'swimming power' can swim at a speed of 0.5 m s^{-1}, what is the effective resistive drag of the water on her body?

6 A tennis player practising 'follow through' in her service manages to keep her racquet in contact with the ball over 0.2 m of her stroke. The ball flies off at 30 m s^{-1}. If the mass of the ball is 100 g, what average force on the ball did she achieve?

Answers **1.** 45 J, 435 J, 1000 J **2.** 735 J, 980 J, 39 J **3.** 2.5 kN, 125 kJ, 14.2 m s^{-1} **4.** 1 in 100 **5.** 400 N **6.** 225 N

YOU HAVE LEARNED

- The kinetic energy of a mass m moving at velocity v is $\frac{1}{2}mv^2$.

- The increase in gravitational potential energy for a mass m lifted through a height h in a gravitational field g is mgh.

- Work done measures energy transferred, and is given by the net force acting multiplied by the component of the displacement in the direction of the force.

- In motion under gravity, with no other forces acting, changes in gravitational potential energy are matched by equal and opposite changes in kinetic energy.

- An object moving at constant velocity against resistive forces loses energy at the same rate as it gains energy from the source which drives it.

- Power expended = rate of doing work = force × velocity (in the direction of the force).

kinetic energy, potential energy, work, power

Summary checkup

✓ **Vectors**

- Displacement, velocity, force and acceleration are all vector quantities, adding 'tip-to-tail'.

- The relative velocity of two objects can be found by subtracting the velocity of one of them from both.

✓ **Kinematics**

- How to construct paths of accelerated bodies step by step.

- Displacements and velocities for uniform acceleration can be predicted using the kinematic equations:

$$v = u + at \qquad s = \frac{u+v}{2}\,t \qquad s = ut + \tfrac{1}{2}at^2 \qquad v^2 = u^2 + 2as$$

- The area under a speed–time graph can give the distance travelled.

- The slope of a speed–time graph can give the acceleration.

- The slope of a distance–time graph can give the speed.

- A projectile in free fall follows a parabolic path.

- Vertical and horizontal components of velocity of a projectile are independent.

✓ **Dynamics**

- Force, mass and acceleration are related by the relationship $\boldsymbol{F} = m\boldsymbol{a}$.

- The gravitational force on an object is proportional to its mass.

- Different masses have the same acceleration g due to gravity.

- The gravitational field g is the gravitational force per unit mass.

✓ **Energy and power**

- The kinetic energy of a mass m moving at velocity v is $\tfrac{1}{2}mv^2$.

- The increase in gravitational potential energy for a mass m lifted through a height h in a gravitational field g is mgh.

- Work done measures energy transferred, and is given by the net force acting multiplied by the component of the displacement in the direction of the force.

- In motion under gravity, with no other forces acting, changes in gravitational potential energy are matched by equal and opposite changes in kinetic energy.

- Power expended = rate of doing work = force × velocity (in the direction of the force).

Questions

Take $g = 9.8$ N kg^{-1} = 9.8 m s^{-2}.

1 This question is about free fall under gravity.

(a) 'Round shot' for use in old muskets was made by dropping molten iron down the inside of a tall tower. The iron formed a sphere and then solidified before falling into cold water. If it took 3 seconds for the iron to solidify how high must the tower be (at least).

(b) You are standing under a bridge and want to know how high it is above you. You (carefully) throw a small stone up at just the right speed to barely reach the bridge. You repeat the measurement several times and find its time of flight (up and down) to be 2.4 s on average. How high is the bridge?

2 Galileo did experiments on the motion of objects. His results were surprisingly accurate considering that he used a water clock to measure time. In a series of experiments he rolled a metal ball down a sloping board, changing the angle of the board to vary the results. The data from one such experiment (in modern units) are as follows: length of board = 3 m, angle of board to the horizontal = 30°, time taken for ball to roll down board = 1.2 s.

(a) What was the (average) acceleration of the ball?

(b) g acts vertically downwards. What is the component of g that acts along the board to produce the acceleration?

(c) From your answers to (a) and (b) what value of g did Galileo obtain? Suggest two reasons why the final result for g is less than the modern value.

3 The following data about a modern motor car are given in a car magazine.

Time to reach 60 mph from 0 mph	9.9 s
Kerb 'weight'	1150 kg
Distance needed to brake from 30 to 0 mph	11.1 m
Distance needed to brake from 70 to 0 mph	63.6 m

(a) Convert the speed data to SI units.

(b) Calculate the acceleration of the car (0 to 60 mph) in m s^{-2}.

(c) What is the average driving force on the car needed to produce this acceleration?

(d) What value of deceleration was required to brake from 30 to 0 mph in a distance of 11.1 m?

(e) What braking force is needed to produce this deceleration?

(f) What value of deceleration was required to brake from 70 to 0 mph in a distance of 63.6 m?

(g) What braking force is needed to produce this deceleration?

(h) Use the data to explain to a learner-driver why it is dangerous to exceed the speed limit of 30 mph in built-up areas.

4 The following data were produced while testing a new type of car.

Time/s	0	0.2	0.4	0.6	0.8	1.0	1.2	1.4	1.6	1.8
Speed/m s^{-1}	0	0.64	1.28	1.92	2.56	3.20	3.84	4.40	5.12	5.76

(a) Draw a speed–time graph for these data.

(b) Describe the motion of the car during this first 1.8 s.

(c) What was the acceleration of the car?

(d) How far did the car travel in the first second?

5 In 'skeet shooting' a clay disc is shot into the air at an angle. In one trial of the system the disc is shot at an angle of 60° to the level ground with a speed of 40 m s^{-1}.

(a) Resolve this starting velocity into its vertical and horizontal components.

(b) Use its upward component of velocity to predict that the disc should stay in the air for about 7 s. Is this likely to be an underestimate or an overestimate? Justify your decision.

(c) How far should the disc have travelled horizontally in this time?

6 The two engines of a 35-seater propeller aircraft can each produce 1000 kW of power for take-off. The take-off mass is 13 000 kg. The take-off speed is 60 m s^{-1}, and the length of runway needed is 600 m. Assume that the aircraft accelerates uniformly up to take-off.

(a) What is its acceleration?

(b) What time does the take-off occupy?

(c) What is the thrust of the propellors on the aircraft?

(d) What is the kinetic energy of the aircraft at take-off?

(e) Calculate the kinetic energy in a different way, from the same data.

(f) What is the power provided to increase the kinetic energy of the aircraft during take-off?

(g) Suggest reasons for the difference between the answer to (f) and the power rating of the engines.

Acknowledgments

The publishers are grateful to the following for permission to reproduce text and images in this book.

1   The poem 'Ultimage Reality' is reproduced by permission of Laurence Pollinger Limited and the Estate of Frieda Lawrence Ravagli

1   (right) Courtesy of Dr Alan Cottenden, UCL

6   (top) [visible light] Digitized Sky Survey, STScI/NASA; [infrared] ESA/ISO/ISOCAM, CEA–Saclay and I F Mirabel *et al*; [radio + infrared] VLA/NRAO (and infrared ISOCAM as stated)

6   (bottom left) AIP Emilio Segrè Visual Archives, Shapley Collection

6   (bottom right) Mullard Radio Astronomy Observatory

8   (left) Courtesy of EML and Lucent Technologies

8   (right) CNES, 1998 Distribution Spot Image/Science Photo Library

10   (top) AIP Emilio Segrè Visual Archives, *Physics Today* Collection

10   (left) Courtesy of Matthias Bohringer and colleagues, University of Lausanne

11   (bottom) NCSA/University of Illinois at Urbana-Champaign

11   (top right) Courtesy of Edward J Groth, University of Princeton

13   Property of AT&T Archives. Reprinted with permission of AT&T

15   AIP Emilio Segrè Visual Archives, See Collection

16   (top) Ground-based image courtesy of Georges Meylan (European Southern Observatory); HST image courtesy of Wide Field and Planetary Camera Team and AURA/STScI/NASA

16   (bottom) Courtesy of Professor M Bertero, University of Genoa

19   Lennart Nilsson/Albert Bonniers Forlag

21   (top) Courtesy of Dolland & Aitchison

29   (top) Courtesy of Professor Richard S Muller, University of California, Berkeley

29   (middle) David Scharf/Science Photo Library

29   (bottom) Reprinted from *Sensors and Actuators* A **34** J Brugger, 193–200, ©1995, with permission from Elsevier Science

30   Courtesy of Dr S T Davies, Director of the Centre for Nanotechnology and Microengineering, School of Engineering, University of Warwick

32   Courtesy of Professor William S Bickel, University of Arizona

33   Courtesy of Oak Ridge National Laboratory, Tennessee

35   (upper) Dr Jeremy Burgess/Science Photo Library

35   (bottom) Courtesy of Professor Robert Wild, University of California, Riverside

38   STC/A Sternberg/Science Photo Library

45   (left) Donna Coveney/MIT

45   (right) Courtesy of Dr A W van Herwaarden and Dr G C M Meijer, Delft University of Technology

47   NASA

48   (upper) Courtesy of Professor P N Bartlett, University of Southampton

48   (bottom) AIP Emilio Segrè Visual Archives, *Physics Today* Collection

49   Courtesy of Professor Julian Gardner, University of Warwick

50   *Microsensors—Principles and Applications*, Julian W Gardner. © John Wiley & Sons Limited. Reproduced with permission

52   Science Museum/Science & Society Picture Library

59   (bottom) The Smithsonian Institution

60   (left) HPL/Bettman Archive

60   (right) Science Museum/Science & Society Picture Library

63   Jeremy Burgess/Science Photo Library

77   Science Museum/Science and Society Picture Library

78   (upper left) Courtesy of the Academic Department of Radiology, University of Sheffield

78   (lower left) Courtesy of Williams Grand Prix Engineering

78   (upper right) Courtesy of Karl Storz Endoscopy (UK) Ltd

78   (middle right) Courtesy of De Beers

78   (bottom right) The Natural History Museum (London)

79   (upper) Photo supplied by Dulas Ltd (specialists in solar medical equipment)

79   (left) The Natural History Museum (London)

80   (right) Drawing by John Sibbick

83   (left) © Dave G Houser/CORBIS

83   (right) Courtesy of Tourism Malaysia

85   Courtesy of BRE

87   Shigio Kogure/Katz Pictures Limited

89   (left) Courtesy of BSI

89   (right) Permission by Instron

90   Dominic Ross, School of Art and Design, Kingston College

92   (left) © The British Museum

92   (right) Pavel Hlava, Czechoslovakia

96   Petit Format/CSI/Science Photo Library

104 (top left) Courtesy of Dr Rodney Cotterill (*The Cambridge Guide to the Material World*)

104 (middle left) Bousfield/BKK/*Science & Society* Picture Library

104 (bottom left) Philippe Plailly/Science Photo Library

104 (top right) Andrew Syred/Science Photo Library

104 (bottom right) Courtesy of Dr Claire Davis, University of Birmingham

107 (bottom left) G Müller, Struers GmbH/Science Photo Library

107 (bottom right) © Daresbury Laboratory

109 (left) Courtesy of RNIB

109 (top right) Photograph supplied courtesy of Pilkington plc

109 (middle right) Courtesy of Auto Windscreens

110 (middle right) Courtesy of the Estate of Michael Ayrton

110 (bottom) Mike McNamee/Science Photo Library

115 (left) Werner Forman Archive, Wallace Collection, London

115 (right) Courtesy of Dr Bernd Kempf, Degussa-Huls AG Dental Division

122 Property of AT&T Archives. Reprinted with permission of AT&T

129 (top) Corbis/Robert Picket

129 (middle) From *On the Surface of Things* by Felice Frankel and George M Whitesides © 1997. Published by Chronicle Books, San Francisco. Used with permission

129 (bottom) From *On the Surface of Things* by Felice Frankel and George M Whitesides © 1997. Published by Chronicle Books, San Francisco. Used with permission

130 (top) Mehau Kulyk/Science Photo Library

130 (middle) Photograph supplied courtesy of Barnabys Picture Library

130 (bottom) Tokyo National Museum

141 David Parker/Science Photo Library

143 AIP Emilio Segrè Visual Archives

146 (bottom) AIP Emilio Segrè Visual Archives

148 (right) Photo Deutsches Museum Munchen

151 (top) University of California

158 (middle) A Rose, *Advances in Biological and Medical Physics* vol 5, no 211 (1957)

158 (bottom) Courtesy of Peter Challis, Harvard Smithsonian Centre for Astrophysics

172 (middle) C Jonsson *Zeitschrift für Physik* **161** 1961

174 Courtesy of the Archives, California Institute of Technology

177 Courtesy of Dr Robert Reid

179 Photo © Anglia Polytechnic University, *Multimedia Motion*, Cambridge Science Media

186 Reproduced from the 1996 1:25 000 Outdoor Leisure Map 38 by permission of Ordnance Survey on behalf of The Controller of Her Majesty's Stationery Office, © Crown Copyright Institute of Physics Publishing Ltd, Dirac House,Temple Back, Bristol BS1 6BE; Licence No. MC 029528

190 (middle column) © Anglia Polytechnic University, *Multimedia Motion*, Cambridge Science Media

190 (right) John Sandford/Science Photo Library

195 (top) Oxford Scientific Films

195 (middle) Michael Dunning

195 (bottom) Gary M Prior/Allsport

200 Professor Harold Edgerton/Science Photo Library

205 Corbis/Ronnen Eshel

207 (left) Mike Hewitt/Allsport

218 National Railway Museum/*Science & Society*

Every effort has been made to trace the holders of copyright. If any have been overlooked, the Publishers will be pleased to make the necessary amendments at the earliest opportunity.

Answers to Questions

Chapter 1

1 a +200 D, b no numerical answer, c 1.23×10^6 pixels,
d 36 mm × 27 mm, e 0.03 mm × 0.03 mm, f yes, g 3.7 Mb

2 no numerical answer

3 c +40 D

4 no numerical answer

5 no numerical answer

6 a 10 nm

Chapter 2

1 a 0.5 mAV^{-1}, b 2 kΩ, c 2 mAV^{-1}, d 4 kΩ, e e.g. 250 mm × 40 mm,
f e.g. 125 mm × 10 mm

2 no numerical answer

3 a 1.25×10^{16} s^{-1}, b 1.6×10^{-15} J, c 20 W, d 100 electrons

4 a no numerical answer, b no numerical answer, c (i) 300 K, (ii) 320 K,
(iii) 0.1 V

5 no numerical answer

Chapter 3

1 a no numerical answer, b sampling frequency = 2 × highest frequency,
c 700×10^3 bits s^{-1}, d 1.4 MHz, e no numerical answer, f 9×10^{-4} mm

2 no numerical answer

3 a 8 kbits s^{-1}, b yes, c 10 000 pixels, each 1 mm × 1 mm

4 a 0.333 m−0.375 m, b 8000 s^{-1}, c 256 levels, d 64 000 bits s^{-1},
e 64 kHz, f 1500 lines

5 no numerical answer

Chapter 4

1 no numerical answer

2 B > A more force, C > B longer, D > C thinner, E > D lower modulus

3 a no numerical answer, b 4000 N

4 a 3×10^{-6} m^2; 1 mm, b 600 N, c 1 mm, d no numerical answer

5 7.6 mm

6 a 60°, b 35.3°, c 2.0×10^8 m s^{-1}

7 no numerical answer

8 a Nichrome, b 1.1×10^{-6} Ω m

9 100 times better

Chapter 5

1 a metals, b semiconductors, c polymers, d metals, e polymers,
f ceramics

2 no numerical answer

3 a about 10 kN, b incisors: perhaps 1 kN; molars much more,
c a few newtons, d 20 kN

4 no numerical answer

5 no numerical answer

6 no numerical answer

7 no numerical answer

Chapter 6

1 no numerical answer

2 a 500 wavelengths, b yes, 6 mm, c same, d reduced to 2 mm,
e slit separation 15 m, f no numerical answer, g 0.2 m

3 a 15×10^9 rotations s^{-1}, b 0.02 m, c same, d no numerical answer,
e 10 m

4 no numerical answer

5 no numerical answer

6 no numerical answer

Chapter 7

1 a yes, b yes, c no numerical answer, d no numerical answer

2

Photons

frequency of phasor rotation/Hz	energy	wavelength
1.0×10^{15}	6.6×10^{-19} J	(300 nm)
(300×10^9)	2.0×10^{-22} J	1 mm
2.4×10^{14}	$(1.6 \times 10^{-19}$ J$)$	1.25 µm

Electrons

frequency of phasor rotation/Hz	energy	wavelength
2.4×10^{15}	(10 eV)	3.8×10^{-10} m
3.6×10^{12}	2.4×10^{-21} J	(10 nm)
(10^{16})	6.6×10^{-18} J	1.9×10^{-10} m

3 no numerical answer

4 no numerical answer

5 a 3×10^{11} mm s^{-1}, b 1 ns, c 1 ns, d 5×10^{-11} s (0.05 ns),
e 2×10^{11} mm s^{-1}, f 1.5

Chapter 8

1 a 30 m, b 0 m s^{-1}, c 25 m s^{-1}, d 12.5 m s^{-1}, (i) same, (ii) bigger,
(iii) same

2 no numerical answer

3 a no numerical answer, b 800 m, c 3.5 km, d no numerical answer

4 a 1.3 m s^{-1} at 67° to bank, b 8.3 s, c upstream at 65° to bank,
d 9.2 s

5 a, b, c no numerical answer, d (i) 38.5 m, (ii) 120 m

Chapter 9

1 a 44 m, b 7.1 m

2 a 4.2 m s^{-2}, b 4.9 m s^{-2}, c 8.4 m s^{-2}

3 a 13.4 m s^{-1}, 26.8 m s^{-1}, 31.3 m s^{-1}, b 2.7 m s^{-2}, c 3.1 kN,
d 8.1 m s^{-2}, e 9300 N, f 7.7 m s^{-2}, g 8860 N, h no numerical answer

4 a, b no numerical answer, c 3.2 m s^{-2}, d 1.6 m

5 a $v_H = 20$ m s^{-1}, $v_V = 34.6$ m s^{-1}, b no numerical answer, c 140 m

6 a 3 m s^{-2}, b 20 s, c 39 kN, d 23 MJ, e no numerical answer,
f 1.17 MW, g no numerical answer

Index

What comes next in *Advancing Physics?*

The A2 Course, leading to a full A-level qualification, covers the following areas of physics:

Advanced Level (A2) Course Content

Rise and Fall of the Clockwork Universe

Models and Rules

10. Creating Models

Simple computational models. Radioactive decay, *RC* circuit, energy stored. Harmonic oscillator, resonance

11. Out into Space

Orbits, circular motion, gravitational field and potential, inverse square law, conservation of momentum

12. Our Place in the Universe

Evidence of origin and evolution of Universe, distance measurement, cosmic redshift, microwave background

Matter in Extremes

13. Matter: Very Simple

Ideal gases, kinetic theory of gases, thermal capacity, conservation of energy, energy *kT*

14. Matter: Very Hot and Cold

Behaviour of matter at high and low temperatures, Boltzmann factor, meaning of temperature

Field and Particle Pictures

Fields

15. Electromagnetic Machines

Transformer, dynamo, motor, power generation and transmission. Flux density and induced emf

16. Charge and Field

Accelerators: Electric field and potential, force on moving charge

Fundamental Particles of Matter

17. Probing Deep into Matter

Scattering, nuclear atom, discrete energy levels, quantum ideas about atoms, particle physics

18. Ionising Radiation and Risk

Radioactive decay processes, nuclear stability, risks from ionising radiations, $E = mc^2$

Advances in Physics

19. Advances in Physics

Case studies of advances in physics and engineering, in which different parts of physics come together to solve problems and to suggest new ideas

Full student support for the A2 Course will be found in the *Advancing Physics A2* Student's Book (ISBN 0 7503 0677 7) and *Advancing Physics A2* Student's CD-ROM (ISBN 0 7503 0738 2) available from Institute of Physics Publishing from Summer 2001.